This is the first book to describe the general features of ZEKE (ZEro Kinetic Energy) spectroscopy, a new high-resolution spectroscopy of molecular ions, neutral short-lived intermediates, and other species.

The author's approach is to use a minimum of equations and large numbers of figures to help the reader towards a basic understanding of the many unique concepts of this new form of spectroscopy and the new spectroscopic information that it provides. Since 1984 ZEKE spectroscopy has matured into a very-high-resolution spectroscopy for the study of cations, anions and, indirectly through these species, of neutral species, including very-short-lived intermediates in chemical reactions. It has even yielded the first direct spectroscopic data on elusive transition states of chemical reactions. It also provides measurement to a very high degree of accuracy and at a resolution three orders of magnitude better than those of other major techniques such as photoelectron spectroscopy. As such, it is able to generate useful new spectroscopic results using a reasonably straightforward experimental set-up.

For positive ions the technique derives its power from the newly discovered existence of certain very-long-lived neutral states of some 50–100 μs duration in a narrow band, some 8 cm^{-1} below each ionic state. These are hidden beneath the strong signal from ions that is always present in normal photo-ionization experiments. Stripping out the signal from these ions produces the ZEKE states, which may be sharpened into a spectroscopic signature of all ion states possible in the system. Such states exist even at much higher energies beyond ionization. For negative ions, this technique looks at the threshold directly since the electron is only weakly bound to the molecular system.

This book will be of interest to anyone interested in the spectroscopy of ions or of neutral species, particularly short-lived neutral species, formed from these ions. It should also be of interest to reaction kineticists interested in the study of reactions involving such highly state-selected species.

ZEKE SPECTROSCOPY

ZEKE SPECTROSCOPY

E. W. SCHLAG
Technische Universität München

CAMBRIDGE
UNIVERSITY PRESS

CAMBRIDGE UNIVERSITY PRESS
Cambridge, New York, Melbourne, Madrid, Cape Town, Singapore, São Paulo

Cambridge University Press
The Edinburgh Building, Cambridge CB2 2RU, UK

Published in the United States of America by Cambridge University Press, New York

www.cambridge.org
Information on this title: www.cambridge.org/9780521581288

First published 1998
This digitally printed first paperback version 2005

A catalogue record for this publication is available from the British Library

Library of Congress Cataloguing in Publication data

Schlag, Edward William, 1932–
ZEKE spectroscopy: the Linnett lectures / E. W. Schlag.
p.   cm.
Includes bibliographical references.
ISBN 0 521 58128 1 (hardcover)
1. ZEKE spectroscopy.   2. Molecular spectra.   3. Ions – Spectra.
4. Chemical kinetics.   I. Title.
QC454.Z44S35   1997
543′.0858–dc21   96-50385 CIP

ISBN-13 978-0-521-58128-8 hardback
ISBN-10 0-521-58128-1 hardback

ISBN-13 978-0-521-67564-2 paperback
ISBN-10 0-521-67564-2 paperback

Contents

Foreword

This book is meant as a primer to the field of ZEKE spectroscopy. Specifically, it attempts to transmit the flavour of the method together with some representative examples of the many applications of this new spectroscopy which are already in existence, by making extensive use of figures and pictures. It is meant to be pleasant reading as an overview of the whys and wherefores of this new spectroscopy – the approach is in line with the historical development, to show the essence of the new method as represented by its evolution, and to present representative examples of the already substantial body of knowledge accumulated in this field leading to a new spectroscopy. For a more formal and rigorous development, the reader is referred to the original literature or to one of the many review articles which now exist on **Z**ero **K**inetic **E**nergy (ZEKE) spectroscopy[1-3] as well as those which are presented on the current list of the World Wide Web (see the end of the book).

The new method was made possible in part by the discovery of ZEKE states; discussion of their origin involves a fascinating sojourn into some new aspects and discoveries of the properties of high-angular-momentum Rydberg molecules, which were not well known before. Only a phenomenological sketch of this physics is given, since this is a different topic in its own right. The usefulness and the applications of this new spectroscopy are legion. It is a new spectroscopy that makes unique use of photoelectrons at threshold. It expands the horizons of our understanding to new structures of import to modern chemistry.

For these lectures I will draw primarily on the work with my group here in Munich. These associates have formed over many years a closely coupled team generating many ideas and experiments that culminated in the groundwork for ZEKE spectroscopy. This cooperative effort has been principally with the group of Dr Müller-Dethlefs, but in addition the groups of

Dr Selzle, Professor Neusser, Dr Boesl, Dr Weinkauf, Dr Held and Dr Baranov have contributed in highly significant but differing ways to the joint effort here in Munich. All of us are joined in the closely intertwined effort of developing this new technique and hence I will refer to this effort in the text naturally as our work. All this work equally well and implicitly reflects many hours of discussions on theory with Professor R. D. Levine in Jerusalem and his associates.

This work, of course, historically was preceded by related work on threshold spectroscopy, starting with a team at Northwestern University principally consisting of my first doctoral student W. Peatman and my two outstanding, then post-doctoral associates T. Baer and P. M. Guyon.

I want to thank Professor David Buckingham, the Chairman of the Linnett Trust, for inviting me to give these lectures from which this book developed. I also want to thank the Master of Sidney-Sussex College, Professor Gabriel Horn, for his kind hospitality during my tenure of a fellowship at Sidney-Sussex.

This book is dedicated to the memory of Professor John Wilfred Linnett, professor of physical chemistry, master of Sidney-Sussex College and Vice-Chancellor of Cambridge University. It is hoped that this short primer will provide much of that which Professor Linnett provided in his two introductory texts on quantum mechanics, namely a useful first guide to students starting in a new field.

Part I
Basics

1

Introduction

At the start it will be useful to cast ZEKE spectroscopy into a general context. The history of chemistry has been punctuated by major developments in spectroscopy. Spectroscopy has given us marvellous insights into molecular shape, structure and constitution. In short, it has taught us the nature of the building blocks of molecules and molecular systems. It has given us an idea of the architecture of chemistry – just as any road going around a city is made easier if one knows the size and shapes of its buildings. In my opinion, obtaining descriptions of such chemical buildings is the function of spectroscopy. Indeed, the structure of molecules serves as an architectural guide-post on our journey through chemistry. Our desire to seek new landmarks in chemistry sparks, of course, our interest in the realization of every new landmark as it comes into view. As chemistry evolves, and I think it is strongly and rapidly evolving, particularly now as one comes to the turn of the century, one is ever on the lookout for new ways of describing the architecture in order to guide us on to new roads for this great journey of chemical understanding. I think that this was once very aptly stated by Lord Todd at the 1985 IUPAC meeting in Manchester. He made the interesting remark that the last century was essentially the century of the chemistry of strong bonds, of tight linkages and the understanding thereof, and, of course, the associated spectroscopy. He felt very strongly at that time that the next century would see the importance of the weak bond, the fluctuating bond and rearrangements. One knows, of course, that this is certainly true, as has been demonstrated in our understanding of biochemical systems, which could not be thought of in terms of hard bonds, but rather in terms of very soft bonds often manifested in the subtle equilibria that make up biochemical transformations. So the subject of weak bonds has certainly emerged from what it was when I was a student of physical chemistry, namely a very small field in which one

3

spoke parenthetically about weak bonded interactions such as van der Waals bonds as an inconsequential subject on the sidelines of physical chemistry. These weak bonds have since become increasingly a core subject of chemistry that is important for much of our new understanding. I think that Lord Todd's comment in 1985 will turn out to have been prescient and more true than most scientists thought at the time. Thus the problem of present day spectroscopy and the applications which are current today can be summarized as a question of the understanding of the architecture of this new chemistry that one is approaching. Here our new questions are not totally answered by our traditional methods, which are Raman spectroscopy, infrared spectroscopy and many other types of spectroscopy. One needs to learn about a series of new systems for architectural structures and landmarks in which these structures are quite different from those one had known hitherto. One example, already mentioned, is that of molecules bonded by van der Waals forces, which had been considered an esoteric area, but has become an emerging subject equal in importance to traditional strong bonding. As a consequence, the understanding of weakly bonded species, with its many consequences for large structures, will move increasingly into the core of our chemical thinking. In particular, one is now starting to understand how these 'soft' systems, which one began to understand in isolated systems, correlate with questions that had arisen in other areas such as in surface spectroscopies e.g. **High-Resolution Electron Energy Loss Spectroscopy (HREELS)**, or **Extended X-Ray Absorption Fine Structure (EXAFS)**. The interlinking of different spectroscopies will perhaps become one of the very important areas one needs to address as one begins to understand the relationship between structures and soft intermediates.

To examine these unusual molecules with soft bonding, and even unconventional neutral molecules and clusters that can be produced only in mixtures, one must consider utilizing mass-selective experimental techniques. What methods are available for neutral molecule spectroscopy, particularly in a mixture, for which mass-selectivity is required? How should one investigate ions, positive or negative? How do charges interact with bonding? How do reactive intermediates appear spectroscopically? Consider for example H_3O^+, NH_4^+ or even more complicated species as typical cases. How can transition states or activated complexes be seen spectroscopically? Transition states in the Polanyi–Wigner–Eyring theory of chemical reactions are characteristically extremely short-lived, typical results suggesting life-times of half a picosecond. How are clusters formed? How do metal–metal and metal–non-metal transitions exist in clusters?

What do surface-bound molecules look like and how do they relate to clusters or to embedded reactions of the solid? How does the whole system interact when it reacts in the presence of the Fermi sea of electrons in the metal? It is in this vein that our initial journey through these new architectural landmarks in soft chemistry is to be launched.

2

Spectroscopy – a historical perspective

This chapter will provide an overview of the various spectroscopies which are of current and topical interest. In the next three chapters, I will naturally address in most detail those spectroscopic experiments that we have carried out in Munich and, in particular, focus on what we have come to call **Z**ero **K**inetic **E**nergy (ZEKE) spectroscopy. The initial development in this direction was work done on the threshold spectroscopy of photoelectrons with W. Peatman at Northwestern University. In Munich, we started again with a variant of this technique. It was the work together with my first group in Munich, K. Müller-Dethlefs and initially, M. Sander, which started our work on the ZEKE variant of this **T**hreshold **P**hotoelectron **S**pectroscopy (TPES).

In a short tour of spectroscopy (Table 2.1), I shall present my personal view of this development. Perhaps it is hard to define the origin of spectroscopy. Perhaps I should go back to Newton, who in 1666 dispersed white light into colours and first coined the word spectrum. Here I choose as a modern date to start with the year 1800, in which Herschel, a musician and astronomer, discerned that the spectrum of light when absorbed by a black body extended to the infrared, i.e. that the sun had a very strong spectrum in this invisible range. Stop number two on my journey is a contribution from Munich. Fraunhofer worked in a laboratory near Munich and was employed to make better optical quality glass. The problem one had in those days was to make glass without bubbles for lenses and for refractive optics. This was very difficult and a laboratory was set up by Utzschneider for this purpose in the Benediktbeuern monastery south of Munich. One can still see the original laboratory there today. Fraunhofer indexed 574 of the famous dark lines found in the sun's spectrum. More than 20 000 of these lines are known today. This pioneering work, published in 1817 in the annals of the Bayerische Akademie der Wissenschaften, was a milestone in spectroscopy.

Table 2.1. *The history of spectroscopy*

Newton 1666
Herschel: infrared 1800
Fraunhofer 1814
Bunsen and Kirchhoff 1859
Rayleigh 1871
Hallwachs 1887 and Einstein 1905
Rydberg 1890
Röntgen 1895
J. J. Thomson 1897
Aston 1912
J. Franck and G. Hertz 1914
Raman and Smekal 1928
Townes and Basov 1954
Maiman 1960
Turner, Terenin and Siegbahn 1962
Herzberg 1971
Bloembergen and Schawlow 1981

The next landmark I would choose is that associated with R. Bunsen and G. Kirchhoff (Fig. 2.1), who might well be referred to as the initiators of the field of chemical application of spectroscopy, i.e. spectral analysis, in which spectra are used for chemical analysis. The cooperation between these two men was very interesting. Although Bunsen was a chemist, he even then recognized the strong interaction between chemistry and physics. Bunsen said that 'a chemist who is not at the same time a physicist is nothing at all' (Ein Chemiker, der nicht gleichzeitig Physiker ist, ist garnichts). Kirchhoff was a physicist who originally worked in Breslau. Bunsen was performing his analysis of systems by looking through filters and observing the system by colour changes. Kirchhoff made the suggestion that, since spectrographs for the dispersion of light were then available, it would be useful to add such an instrument to improve the spectral analysis. So it was that, during a famous evening walk at sunset in Heidelberg on the Philosophenweg, Bunsen convinced Kirchhoff to join him in Heidelberg and abandon his position in Breslau in order to start work on this historic cooperation of physics and chemistry which led to what is now spectral analysis. Their initial application of this spectral method to the mineral springs in the Bad Dürkheimer Mineralbrunnen led to the discovery of the spectra of rubidium and caesium – although the real problem of the Maxquelle was arsenic.

Then, of course, came the early work of Hallwachs on the photoelectric effect, involving the observation of the charging of a metal plate when it is

Fig. 2.1 R. Bunsen and G. Kirchhoff on a walk (courtesy of Deutsches Museum München).

irradiated with UV light, and of Einstein, who interpreted it. At this point, I have to mention the great discovery of the electron by J. J. Thomson in Cambridge. In Cambridge there began the work which not only brought about our understanding of the electron, but also laid the foundation of early mass spectrometry. Spectroscopy continued in many centres, including the well-known work of Rydberg and his famous formula for cataloguing the hydrogen spectrum, Röntgen's original work with X-rays, first 100 years ago in Würzburg and later in Munich, the scattering experiments of Rayleigh utilizing the sun's rays, the determination of the energy levels in the mercury system in 1914 by the Franck–Hertz experiments, the first measurement by Raman employing what is now known as the Raman effect following an early suggestion by Smekal and the encyclopaedic work of G. Herzberg.

Photoelectron Spectroscopy (PES) has an equally rich history (Table 2.2). The three famous schools should be mentioned: those of David Turner in Oxford, Terenin, Kurbatov and Vilesov in Leningrad (now Sankt Petersburg) and Siegbahn in Uppsala. These three groups contributed enormously much to the pioneering work on which many people's understanding of photoelectron spectroscopy is based. Their work has been so important, indeed instrumental, to our understanding of molecular structure that I think much of what is known today about molecular structure is due to this beautiful work which encompassed the vacuum ultraviolet region of molecular orbitals as well as the atom-specific core excitations with X-rays.

The first demonstration of an optical laser, performed by Maiman, used the three-level ruby system. Recent work on lasers was carried out by Townes and Basov, on the basis of which Bloembergen, Schawlow and Hänsch developed atomic laser spectroscopy.

However, of course, spectroscopy provides understanding of systems and of their structures. Here an important individual was August Kekulé (Fig. 2.2), who was one of the early proposers of strange non-rigid structures in the benzene bonds and extended the discussion to the question of the alternating double and single bonds, which already was a departure from the idea of simple bonding. This is a story that continues today with the structure of benzene dimers and trimers.

It has been reported that much of what Kekulé did he invented on a tram in London. Some students decided during a bar-room celebration that what he was really talking about in benzene was that which is shown in Fig. 2.3. These well-known pictures were generated by them in a later spoof article.

To get into current practical examples in spectroscopy, let us consider whither the work of Kekulé leads us today. At the top of Fig. 2.4 one has an excitation spectrum of benzene at normal dye laser resolution. Now, at the resolution of one wavenumber (1 cm^{-1}), it shows the vibrational transitions which are involved in two-photon absorption, this being the absorp-

Table 2.2. *Spectroscopy with photons and electrons*

PIE		(1954)
Photo-ionization efficiency		
Watanabe	Cambridge, Massachusetts	
PES		(1961)
Photoelectron spectroscopy		
Siegbahn	Uppsala	
Turner	Oxford	
Terenin	Leningrad	
TPES		(1969)
Threshold photoelectron spectroscopy		
Peatman	Evanston	
REMPI–MS		(1978)
Resonance-enhanced multiphoton ionization-		
mass spectroscopy		
Boesl	Munich	
Bernstein	Columbia	
Letokhov	Moscow	
RETOF		(1973)
Reflectron time of flight		
Mamyrin	Leningrad	
Boesl	Munich	
ZEKE		(1984)
Zero electron kinetic energy		
Müller-Dethlefs	Munich	
Anion ZEKE		(1989)
Neumark	Berkeley, California	
Boesl	Munich	
European Science Conferences		
Kreuth		(1991)
Giens		(1993)
Lenggries		(1995)

tion of two photons at high laser intensity. In benzene, the electronic $S_1 \leftarrow S_0$ transition is forbidden both for one-photon and for two-photon absorption. The first band shown here is vibrationally allowed in the two-photon spectrum and is induced by the vibration called ν_{14}. This vibration has the same symmetry as the electronic S_1 state, namely b_{2u}; the transition is then allowed when the ν_{14} vibration is excited in S_1. Clearly, two-photon transitions have selection rules different from those for one-photon transitions, in the same way that a Raman transition has selection rules different from

Fig. 2.2 August Kekulé (courtesy of Deutsches Museum, München).

Fig. 2.3 A caricature of the new bonding in benzene according to A. Kekulé
(excerpt from a spoof by his students).

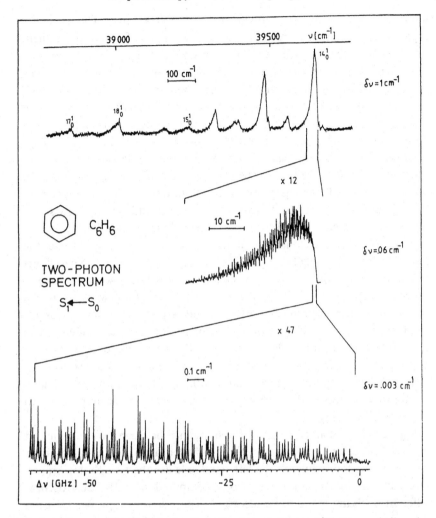

Fig. 2.4 The use of high resolution in molecular spectroscopy[4]: (a) normal vibroni-cresolution, (b) high resolution up to the Doppler limit and (c) sub-Doppler resolution, showing individual rotations.

those for an infrared transition. There is a very close analogy between the two spectroscopies, as Maria Göppert-Mayer showed many years ago[5,6].

At the top of Fig. 2.4 is a spectrum that we have reported earlier in our work. It is a spectrum showing just the vibrations in the excited states. One could ask what this first band would really look like if one could have slightly better resolution. The middle of Fig. 2.4 shows this system with better resolution, which is about the best modern commercial pulsed dye lasers can achieve, at the Doppler limit, some 0.06 cm^{-1}. This is then a typical dye laser spectrum. This spectrum is a bit of a disappointment because where there was previously a nice smooth line in the top spectrum, now there is some apparent 'noise'. However, this is not noise; it is really the inception of rotational structure since these rotational levels are much more closely spaced than the vibrational levels. Hence, one can see a lot of this rotational structure. To prove this point, we set out to achieve sub-Doppler molecular spectroscopy. This was the first example of a sub-Doppler electronic spectrum of an excited state of a molecule. As an example, if one looks at the very beginning of the onset as it arises near the origin of the band and expands it, one obtains a highly resolved spectrum (the bottom part of Fig. 2.4), which starts at zero and is very richly structured, but has a clear base line. One is now below the Doppler width and one can see single rotational lines in this particular vibrational band at the resolution of a few megahertz (1 MHz$=3\times10^{-5}$ cm^{-1}). We did experiments of this type some time ago and in the meantime it has become possible to generate such spectra quite readily. It is quite straightforward to fit this spectrum to an appropriate Hamiltonian and, in fact, the fitting and assignment of such spectra has become an exercise for research students. The great difficulty was not analysis, but rather the generation of these spectra in the first place since the oscillator strength f is so low. These transitions do not have $f=1$ but rather have about $f=10^{-5}$ due to dilution over so many ro-vibrational states, which made the experiment almost impossible to perform successfully.

Figure 2.5 is intended to show the difference between two-photon and one-photon spectroscopy again. On the right-hand side are the typical one-photon transitions while on the left-hand side are the two-photon transitions. Just as in Raman spectroscopy compared with IR spectroscopy, different levels are found to be accessible. These states are probed by using them as resonant intermediate states in a **Resonant Multiphoton Ionization** (REMPI) experiment, as was done in a scanning single experiment in our laboratory by Boesl *et al.*[7] and also by Zandee *et al.*[8] in 1978 in the group of Bernstein at Columbia, who also coined the acronym. This difference in transitions in the molecule is quite normal given the high symmetry of benzene. Below approximately 2000 cm^{-1} of excitation energy, one has modes that are essentially isolated. This realm is often referred to as mode-specific excitation. At higher vibrational energies above the ground vibrational state, all modes become mixed by anharmonic interactions, Fermi resonances, etc. This is sometimes referred to as the channel three region[9]. This is really not a very useful nomenclature. We now know that the reason

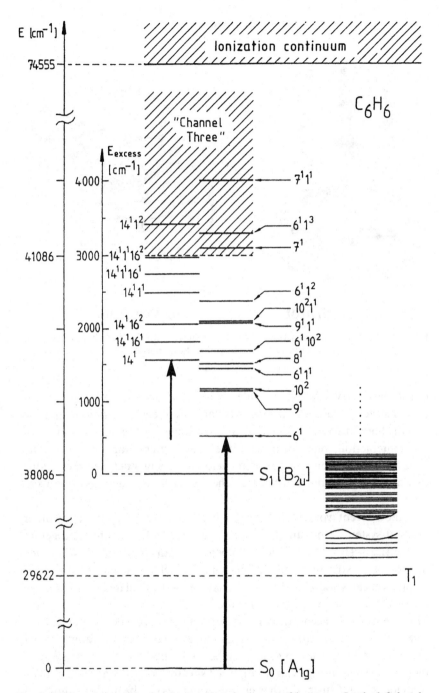

Fig. 2.5 Vibronic levels in the electronic-excited state of benzene. On the left-hand side are the two-photon active modes; on the right-hand side are the one-photon active modes. Above 3000 cm⁻¹ strong mixing occurs in the 'channel three' region.

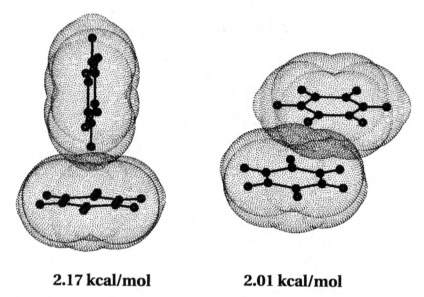

2.17 kcal/mol   **2.01 kcal/mol**

Fig. 2.6 Conformational isomers of the benzene dimer together with their respective calculated stabilization energies[12].

is just that the level density becomes so high that everything is just one statistical system. One can no longer isolate any specific vibrations and the natural linewidths are wider than the level spacing. In the beginning of these interactions one can observe these level interactions in the S_1 state directly as a function of rotational energy. By using a detailed theory due to Marcus *et al.*[10,11] one can calculate these complex interactions quite accurately.

In more recent work, we have expanded this structural approach to other kinds of systems. One can ask how two benzene molecules are put together geometrically (Fig. 2.6). Indeed, two benzenes can couple in two ways. One finds a T-type structure and a displaced sandwich conformer, which, from the theoretical work of P. Hobza[13–15], have similar calculated energies using a CCSD(T) method program. This leads to their facile interconversion. More recently, benzene trimers have been analysed spectroscopically[16]. Again, two main structures with a triangular or a double-T arrangement of the individual benzene molecules can be considered theoretically. Some of these molecular clusters can be analysed spectroscopically and their structures identified. Clusters with even higher masses can be made as shown in a mass spectrum (Fig. 2.7). Here it is seen that the mass spectrum in a supersonic jet expansion of benzene in helium can produce very large clusters.

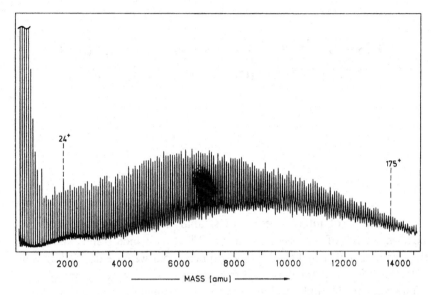

Fig. 2.7 The cluster mass distribution in a benzene jet expansion[17].

For example, 175 benzene molecules can be bound together. Thus, one can generate a very rich spectrum from such clusters, as has now been done in many different laboratories. The problem today is really not to generate a complex cluster spectrum, but rather to study one of these complexes of a given mass and to make sure that just one defined cluster is being examined. This is really only practical up to $n=3$ or 4, rarely above, unless very poor resolution is acceptable. The days of just simply massively producing clusters are past and yet the detailed spectroscopic work on a specified cluster required to determine its structure is quite difficult. The principal experimental difficulty is due to the 'raining down' from higher masses. Typically, a cluster system absorbs spectrally in a cluster at mass $n+1$ times the monomer mass, which then can decompose by ejecting one monomer and shows up at mass n times the monomer mass in the mass spectrometer. This process leads to the mistaken assignment of absorption at mass n times the monomer mass. Further complications can come from ZEKE electron cross talk (*vide infra*).

3

Mass analysis and resonance enhanced multiphoton ionization (REMPI)

The next topic of interest here is mass analysis. Here one can display a cartoon version of a mass spectrometer (Fig. 3.1), which I think shows all the essential elements. The mass spectrometry that we have decided on is a reflecting time-of-flight mass spectrometry suggested many years ago by Mamyrin and Shmikk in Leningrad[18]. The special features were long hidden in the literature. We revived this technique employing high-resolution laser applications and married it to REMPI[19]. Some very interesting new experimental properties of this design have become evident. Basically, the time-of-flight method is conceptually the simplest form of mass spectrometry. You produce an ion at one point, you accelerate it in one direction and you time how long it takes to reach a detector. The small masses arrive at the detector faster than the large masses, whose flight can take a long time. This is basically an old technique. There were commercial instruments built using the linear version of the technique many years ago, but they never worked very well. In particular, the mass resolution was quite low.

The suggestion that Mamyrin and Shmikk[18] made is very simple. They said that the errors due to different positions of production of the ions that cause bad resolution originate in the ionization region and arise from starting at different potential energies. These errors can be corrected by simply reflecting the whole system of ions back on itself with an electrostatic mirror, ideally a soft mirror much like an electrostatic pillow to permit different turn-around points for the various energies. So one need just fold the path. With the energy-dependent transit time in the electrostatic mirror, it is possible to compensate for most of the errors produced in the source region of the instrument. Early realizations of this technique were for electron-impact applications which lead to lower resolution.

A particularly powerful form of mass spectrometry results if photo-

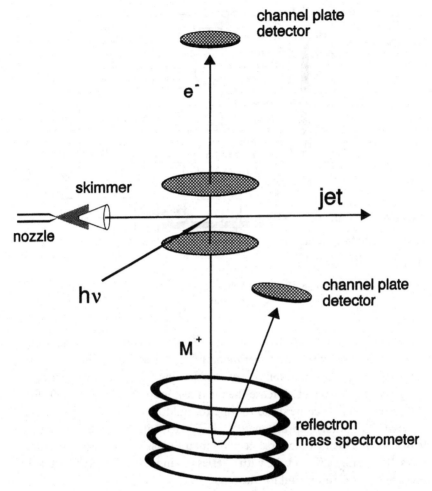

Fig. 3.1 A symbolic representation of the jet experiment with a reflectron mass spectrometer and electron collection.

ionization proceeds as a two-step process via a resonant intermediate state. The molecule- specific character of the intermediate state adds a new, additional dimensionality to mass spectrometry, referred to as REMPI (resonant enhanced multi-photon ionization). This technique is now in use in most laser mass spectrometric laboratories across the world. The original definitive experiments were performed in 1978 by Boesl et al.[7] and independently by Zandee et al.[8] at Columbia. In particular Boesl et al.[20] could show that, by using the REMPI method, the mass spectrum and absorption spectrum of pure ^{13}C-benzene could be produced even though the

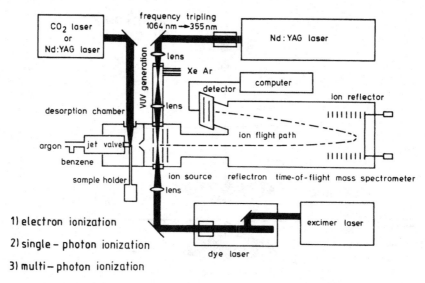

Fig. 3.2 The experimental set-up of two-colour photo-ionization with laser vaporization.

sample was a natural mixture with 6% ^{13}C. The technique was also suggested by Letokhov[21]. The experimental REMPI set-up is shown in Fig. 3.2. Laser excitation gave some very encouraging results. In this particular design, we could in fact increase the mass resolution to about 20 000 (Fig. 3.3). Now it has become quite routine to build these so-called **Reflectron Time-of-Flight** mass spectrometers (RETOF) with a resolution of 20 000–30 000. This is, of course, a poor man's mass spectrometer in many ways. It is simply a flight tube with a series of rings at the end (not grids) to provide a soft turn-around for the ions.

In the course of this work, we discovered[22] another very important property of reflectrons. This application is based on the fact that molecular or cluster ions carry fragments along their flight from the acceleration region to the detector. Of interest here are fragmentations in two distinct flight regions: (i) in the extraction field region where the ions are accelerated and (ii) in the free-flight region where ions with different final velocities separate spatially. If a molecule decomposes in the accelerating region where it is still subject to Newton's laws of acceleration, then, of course, the time of arrival will be smeared out depending on where in this region the mass breaks into fragments. Hence, ions are accelerated in different ways (the top of Fig. 3.4), producing a broad peak which gives an indirect picture of reactivity. Once the ions have left this acceleration region and passed into

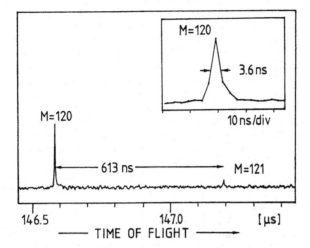

Fig. 3.3 Mass resolution *R* (50%) >20 000 at mass 120 obtained in the reflectron mass spectrometer with trimethylbenzene[19].

the drift region, the **Linear Time-of-Flight (LINTOF)** system is blind to their further decomposition. The system is now put in motion and, even if the molecular ion falls apart in the drift region, all fragment parts will still arrive at a distant wall at exactly the same time as the ion of the original mass does. In the uncorrected spectrum at the top of Fig. 3.4, the fragment peak is hidden under the parent peak. In the accelerating region of this spectrum, peak distortion due to fragmentation occurs. Most of the kinetic work in mass spectrometry is done by analysing this peak distortion. What Boesl *et al.* discovered[19] was that, if they tuned the electric potential of the reflector rather gently, as a partial electron pillow, then they could separate this daughter peak from the parent peak into a third peak (the second spectrum of Fig. 3.4) due to decomposition in the drift region. Generally, this is just a so-called metastable peak, but with the additional advantage that, depending on how hard or how soft you tune the reflector, you can successively shift this peak away from its parent. Hence, you always know what the origin of the daughter is, i.e. from which parent it came. It is a very interesting way of actually looking at parent–daughter–parent–daughter–parent–daughter-type hierarchies and precisely measuring their lifetimes. This is quite useful for understanding the kinetics of very complex systems. In Fig. 3.5, we show this parent–daughter relationship for the dimer of *p*-difluorobenzene. This can be used to disentangle very complex sets of mass spectra. For example, Fig. 3.6 shows a series of these relationships for clusters of benzene with *p*-difluorobenzene. Fine tuning of the reflectron volt-

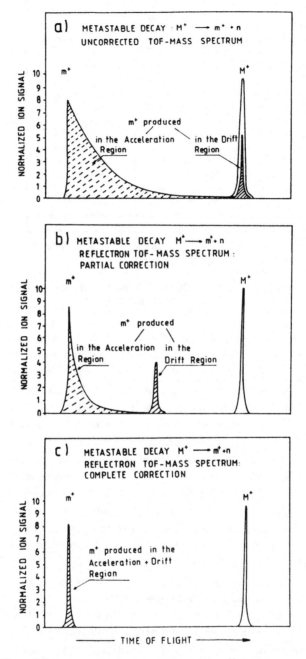

Fig. 3.4 Three different operating modes of a reflectron[23]. (a) No correction, i.e. the same as is found in a linear TOF-MS. All the ions which decompose in the drift region are counted as parent ions. (b) Some correction, showing a new metastable peak from the drift region. This is the operational mode of choice. Note that the middle peak can be shifted, which demonstrates its parentage. (c) Complete correction, adding products from the initial acceleration region and the drift region.

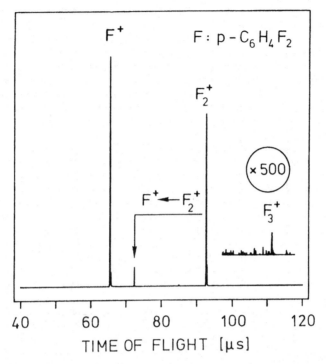

Fig. 3.5 Reflectron kinetics of *p*-difluorobenzene demonstrating parent–daughter relationships. Note the two F^+ peaks due to the two regions of decomposition[24].

ages gives a mass spectrum that exhibits all metastable decay channels as well as the parentage of every fragment mass and even a hierarchy of parentages.

Our most recent application of this method considered some practical aspects. We are engaged in the trace analysis of pollutants in automobile exhaust by multiphoton mass spectrometry. The separate trace components are laser-excited in a **Resonance-Enhanced Multiphoton Ionization (REMPI)** experiment using a reflectron mass spectrometer for mass analysis in environmental systems. By going via a resonant intermediate state with REMPI one has a new selectivity for the system. In the first step, a UV photon is absorbed, as in an ordinary UV spectrum, which is molecule-specific. Then one ionizes with a second photon. Note that the REMPI method is species-selective by virtue of being spectroscopically selective. As an example, we can show the pure carbon monoxide mass spectrum in an air sample in which nitrogen is the major species, which appears at the same mass (Fig. 3.7). The most important point of relevance here is the incidental advantage of the rapid scan of the RETOF mass spectrum. This now

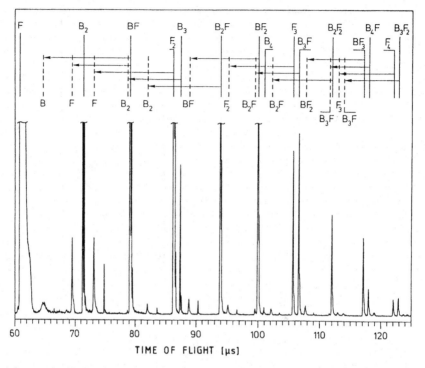

Fig. 3.6 Mixed benzene–fluorobenzene clusters together with their parent–
daughter peaks. From this kinetic rates can be determined[24].

allows one to record a complete mass spectrum for every laser pulse, some
50 mass spectra per second. This allows us to do chemical analysis in real
time, which is a most important feature for environmental monitoring.

Another application is laser tandem mass spectroscopy whereby two
mass spectrometers are operated in tandem (Fig. 3.8). Within the first TOF
mass spectrometer, the different ions are separated into their various mass
components with a linear mass spectrometer utilizing second-order space
focusing. After this a second laser excites these ions in a mass-selective way
by focusing this laser into the space focus and delaying it in time until the
molecular ions of interest pass the space focus. The laser excitation results
in dissociation of the ions. There secondary fragments are analysed by the
second mass spectrometer of reflectron type. Again, the original ions are
formed by REMPI. This laser tandem mass spectrometry allows analysis
of the structure or type of parent molecules, similarly to conventional mass
spectrometry, but with much higher speed, much simpler construction and
considerably larger variability.

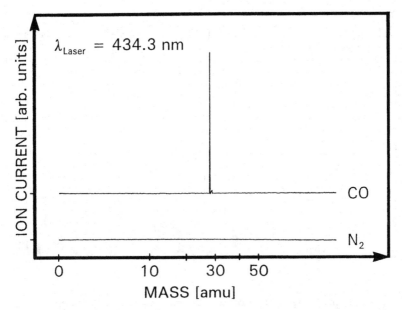

Fig. 3.7 A demonstration of REMPI selectivity. Only the CO peak is seen in an exhaust spectrum in air even though CO and N_2 have the same mass in the MS.

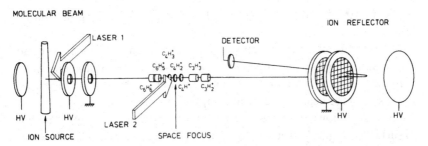

Fig. 3.8 A laser tandem mass spectrometer. The first MS is a linear TOF-MS of resolution 1000, the second is a reflectron. The first MS separates the initial fragments. The second MS provides decomposition kinetics of these initial fragments[19,25].

REMPI state selectivity is also important for defining the states for the final transition on the way to ionization. Hence, the selection rules from the S_1 state are projected into the ionic state. A well-defined level in the S_1 state can be selected with the first photon. This technique can, in fact, be more strongly selective than can possibly be achieved by starting in the ground state, even with extreme cooling. To show what this involves and the use-

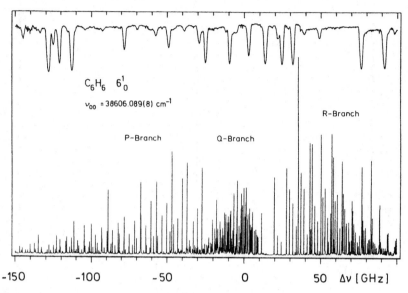

Fig. 3.9 A REMPI high-resolution sub-Doppler spectrum of benzene; note the gigahertz resolution[26].

fulness of REMPI spectroscopy, an early spectrum of ours is shown in Fig. 3.9.

One particular use for the REMPI method is in measurement of bond dissociation energies of complexes. As one increases the photon energy of the ionization step, complexes fall apart and the decomposition increases. This is shown for benzene–toluene complexes in Fig. 3.10.

This dissociation can be explained as decomposition on the upper potential surface in Fig. 3.11, leading to the appearance potential (AP). The ionization potentials of the monomer and the dimer can be measured separately in a similar experiment. From this, and the simple cycle in Fig. 3.11, one can obtain the dissociation energies for the ground-state dimer and the dimer ion. Some values are given in Table 3.1.

REMPI also has very important applications in ZEKE spectroscopy as well as in modern versions of PES.

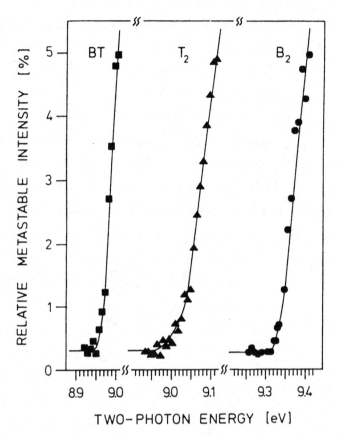

Fig. 3.10 The decomposition spectrum of benzene–toluene complexes as a function of excess energy[27].

Table 3.1. *Dissociation energies of clusters in the ground and ionic states from the Born-Haber cycle*[24,28,29]

	IP (eV)	AP (eV)	D_0 (meV)	E_0 (meV)
B_2	8.65±0.01	9.31±0.01	70±10	660±20
T_2	8.34±0.01	8.97±0.01	150±10	630±20
F_2	8.87±0.02	9.25±0.02	90±20	380±40
BF	8.75±0.02	9.24±0.02	80±20	490±40
BT	8.42±0.01	8.95±0.01	130±10	530±20
BC	9.12±0.02	9.32±0.02	80±20	200±40
BAr	9.222±0.001	9.260±0.005	17±5	38±5
BN_2	9.227±0.002	9.283±0.003	40±3	56±5
B_3	8.58±0.02	8.85±0.02	200±20	270±40
B_4	8.55±0.02	8.68±0.02	100±20	130±40
B_5	8.50±0.02	≤8.61	≤60	≤110

Note:
B, benzene; T, toluene; F, *p*-difluorobenzene; C, cyclohexane.

$$D_0 = AP - IP(B)$$

$$E_0 = AP - IP(B_2)$$

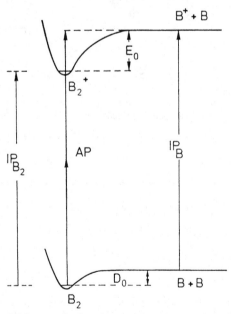

Fig. 3.11 A Born–Haber cycle for determining bond dissociation energies of complexes in the ion and also in the ground state. It requires the measurement of the appearance potential (AP) of fragments of the complex and the measurement of the ionization potentials of the complex and of the monomer[24].

4

Photoelectron spectroscopy

In this chapter the general question of measuring spectroscopic results by detecting the photoelectron emitted after the photon has been absorbed is addressed. Basically, if the photon energy exceeds the ionization potential (IP), better called the ionization energy, the energy suffices to emit an electron and consequently leave an ion behind. There is a multitude of molecular ionic states above the ionization limit, each being connected to a Rydberg ladder, which converges at high energy to the ionic state. There are then as many Rydberg ladders as there are ionic states. Each ladder originates below the ionization potential, but has a point of convergence above, maybe high into the continuum. The photoelectron is not only emitted at the convergence of the ionic state, but also at energies above this. Once the ionization channel has been turned on it will generally remain turned on at higher energies. This is so since the ion and electron pair produced in photo-ionization continue to absorb at higher energies by increasing the relative kinetic energy of the two fragments. As the next transition is formed this simply adds to the signal. In addition there are auto-ionization resonances in the continuum, which are high-energy neutral states that can auto-ionize at defined energies, hence producing additional sharp lines.

This opens up a series of possibilities for measuring such ionic states. One could just scan the total current with energy, leading to a scan of the photo-ionization efficiency (PIE) of Watanabe. This is a staircase function, at least at the start. A recent version of this type of scan is shown in Fig. 4.1.

Alternatively, since the cross section survives the excess energy, one could use light of a single frequency, high-energy Vacuum UV radiation (VUV). This is referred to as **Photoelectron Spectroscopy (PES)**. The VUV wavelength employed is typically the helium 21 eV line from a helium lamp, which emits just a single line at 584 Å. This energy is typically above the

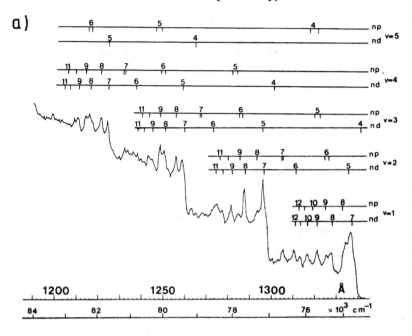

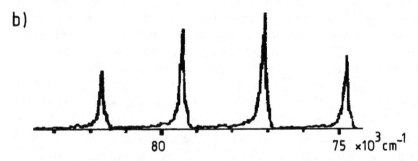

Fig. 4.1 (a) The staircase function typical of PIE together with a Rydber background for the case of nitric oxide[30]. (b) The nitric oxide ZEKE spectrum for comparison.

ionization potential of all molecular systems. The molecules emit electrons with varying energies corresponding to the final molecular energy levels in the ion. The kinetic energy of these emitted electrons can be measured using an electron monochromator. By knowing this energy, one can iden-tify the energy of these molecular levels. This can be done with the helium resonance line in an overshoot technique or with specific laser wavelengths in a multiphoton overshoot. There are many ways one can excite the system. In general it does not matter which energy is used, unless special resonances are excited, but smaller excess energies generally provide better accuracy and resolution for the electrons. In particular, one can also go to higher energies and use the He II line or one can go up to the X-ray region for inner shell information. This depends on the states of interest. The experimental problem reduces to a measurement of electron energies. Yet the measurement of electron energies, for various reasons that are only understood in general, is a very difficult undertaking. It is very hard to measure energies with great accuracy, especially in an absolute sense. This typically requires the constant presence of a calibration gas. More difficult is the question of energy resolution. The resolving power in PES typically is of the order of 10 meV, this being some 80–100 cm^{-1}, although sometimes values as low as 3–5 meV have been reported[31] and values down to 1 meV are now typical in EELS. The absolute accuracy can drift simply because you have to pull the electrons past some type of a surface, an electron monochromator or other instrument. The special coatings recommended to address this problem resemble black magic, since they are irreproducible and produce a drift with time. All these surface potentials are experienced by the electrons. In addition, we have external stray fields that often come into the apparatus despite our best efforts. This distorts the spectra energetically and limits the resolution of all spectrometers. This situation exists even after some 30 years of work on photoelectron spectrometers. There must be 10–20 variants of such instruments in existence by now. They all have one thing in common. They are equally limited in their resolution and their drift in the absolute energy axis.

5

Threshold spectroscopy

In our own early work a new *Ansatz* was considered. Maybe we should attack this whole problem from the opposite end by tuning the photon source and developing a special detector for electrons at the spectroscopic threshold. How was the photon source tuned in those early years? There was no laser, so we employed a high-pressure argon arc with a vacuum UV monochromator and confocal Schmidt–Cassegrain optics. When one reaches the ionization potential, the molecule has absorbed all the energy necessary to produce this particular ion and, hence, there is nothing left over for the electron. Hence, if one measured the electron energy, one should find that it has zero kinetic energy. This is the simplest idea of a threshold detector. One has only to determine when such threshold electrons are made. The apparently most obvious way of thinking about this experiment would be to find out whether one can connect it to an electron monochromator that is tuned to zero energy and thus is capable of measuring energy peaks at zero energy. This approach, however, has the same faults as the original technique, since it will have the same poor resolution and energy drift and hence will improve neither resolution nor accuracy. This was shown in early experiments by Villarejo *et al.*[32]. Actually, these experiments were not carried out at zero energy but at a draw-out field of 3–6 eV.

Thus, we considered a new *Ansatz* for measurement and hence went to an exact threshold detection method based on a different property of the threshold electrons, not their energy. We referred to this as the steradiancy method, since it exploits the steradiant properties of the source. That is to say, one should note that threshold electrons are unique in that they can be emitted without angular divergence of the electrons from the source. This is unique and applies only to electrons of exactly zero kinetic energy. A very important additional point to realize is that this technique yields energies

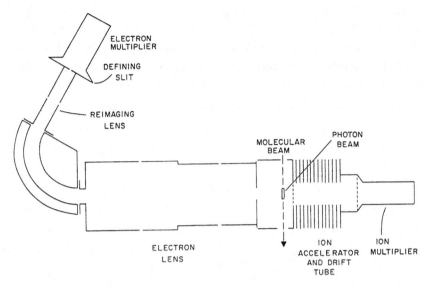

Fig. 5.1 Note the apparatus functions as a steradiancy analyser and as a discriminator against straight-through electrons. The ions measured on the right-hand side in coincidence are thus state-selected[34].

for the spectroscopic transitions with absolute accuracy – as good as the absolute calibration of the light source. To put it in a very simplified picture, a molecular ion at threshold is unique in that it just pops out the electron without an initial velocity. One can produce these electrons, line them up and, with a slight draw-out field, thread all of them through a hole in the exit of a flight tube, with no angular dispersion. If they have any excess energy, they will have cross components transverse to the draw-out field and hence possess angular dispersion. However, this cannot falsify the energy position of their peak. These cross components will prevent these hot electrons from reaching the small hole in the exit plate. They will simply diverge and hit the wall of the tube. This process works best for the worst optical acceptance angle. In this simple view, one wants to build the world's worst spectrometer with the smallest acceptance angle and the worst f number and let this high-f-number optics be an easy way of detecting electrons with energies just at threshold. There is, of course, a small background signal due to hot electrons accidentally emitted in the flight direction. These can be suppressed by employing a time-gated detection scheme[33] or a curved detector (Fig. 5.1) as done by Peatman[34].

As one reaches a particular molecular energy level in the ion, one produces these threshold electrons since all the energy is consumed in producing the state. As one increases the energy again beyond this state, there will only be hot electrons, albeit with strong intensity. These do not make it through the drift tube; hence, the signal disappears. Since this is a two-particle problem, producing ions and electrons at all energies, a threshold signal is not observed at all energies, but only at the photon energies leading to threshold electrons.

6

Zero kinetic energy (ZEKE) spectroscopy – an introduction

In the ZEKE technique we do something quite in addition to measuring the total electron or ion signal as a function of photon energy or even measuring these at threshold. This signal as a function of photon energy (PIE) shows little if any structure of interest. The first challenge is to unveil the structure that one now knows lies beneath this total signal by filtering out special states that we term ZEKE states. The second goal is to do this with laser-limited resolution. These specially prepared highly resolved ZEKE states give the direct high-resolution signature of all the ionic states in the continuum and hence afford a new spectroscopy.

In a near-zero field, or as close to a zero field as one can get, one excites the system again at a particular energy in the vicinity of a threshold to an energy level in the ion. Then one does nothing. One simply waits for several microseconds (Fig. 6.1).

At this point, the whole cloud of charged particles blows up. The near-zero kinetic energy electron created from excitation just slightly above a threshold will be expanded away from the true ZEKE electrons. The arrival time differences at the end of this flight tube, of course, will be greatly amplified even for the in-line components. This is analogous to a time-to-amplitude converter. One is delaying the system after production of the charged cloud and, by delaying the system, one greatly increases the timing accuracy by pulling apart all of the electrons with close-lying kinetic energies. An experimental realization of this method is shown in Fig. 6.2. How does this relate to a total current versus photon-energy scan (PIE)? Let us go back to this conventional excitation spectrum (Fig. 4.1). Suppose that I simply scanned the signal through the ionization to higher and higher energies. I go more and more into the vacuum ultraviolet and observe the onset current. This simply means that I have reached ionization. Eventually there are some squiggles, Rydberg series and various other transitions, auto-

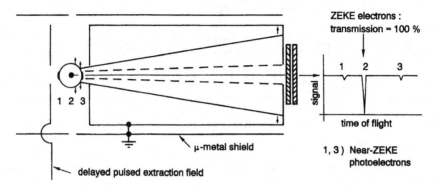

Fig. 6.1 The principle of steradiancy selection of threshold electrons. The improvement in steradiancy detection obtained by delaying the draw-out pulse is also demonstrated.

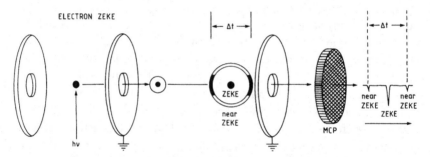

Fig. 6.2 A ZEKE electron flight tube. Here the increasing size of the ZEKE cloud due to near-ZEKE electrons, with some perpendicular velocity component, is shown.

ionization resonances and the like. Then the first vibration appears, then comes the second vibration, the third vibration and so on. The difficulty is that the ion signal increases as one increases the energy in a cumulative way. The reason is simply that production of photoelectrons and photo-ions is a two-particle problem and the energy conservation laws can always be fulfilled by simply having the particles separate faster with the excess energy going into kinetic energy. The photo-ionization efficiency curve, therefore, is at best like an increasing staircase function. This would still be easy to analyse, but, when high energies are reached, even the stairs are no longer visible and the spectrum appears to be full of complicated resonances, which in some cases can be identified as being due to multiple Rydberg series. This complication becomes so severe that only a few systems have yielded to a full spectroscopic analysis of the Rydberg series in this high-energy region. What has been done with the ZEKE energy detection scheme is that one has converted the photo-ionization efficiency spectrum, which is a total current spectrum, to one looking only at the zero kinetic energy component of the total current. This produces again a normal type of spectrum of peaks at energy states and, for the first time, a scan through the states. In this way we unveil the sharp ZEKE peaks lying beneath even a smooth PIE scan. To facilitate further exposition of the physical principles underlying the ZEKE spectroscopic technique, it is convenient to distinguish, from the beginning, between ZEKE spectroscopy of positive ions M^+ and that of anions M^-. The first provides an energy spectrum of the cation M^+ produced by photoexcitation and subsequent ionization of the neutral precursor M. The second yields the energy levels of the neutral molecule M obtained by the photodetachment of the electron from the molecular anion M^-. In the second case, the delayed detection signal is made up from ZEKE electrons that are free electrons liberated from the anion M^- at zero kinetic energy. The first step towards understanding the ZEKE spectroscopy of neutral precursors was made when it was discovered[35] that the delayed detection signal is due to the weakly bound very-high-n Rydberg states which live long enough to survive the delay and which can be readily ionized by even a small draw-out electric field. In the rest of this chapter we discuss the first method, that based on Rydberg states.

6.1 Apparatus

Absorption in a Rydberg manifold (Fig. 6.3) takes place to very high values of n, but to low values of the azimuthal quantum number ℓ. As ℓ increases the electron goes into ever larger orbits around the ionic core. The closest approach to this core goes as $\frac{1}{2}\ell(\ell+1)$, hence the higher ℓ the more the electron stays away from the core. In principle ℓ can increase up to $n-1$ in value.

For Rydberg states in the range $n=100$–300 the state density becomes very high so that small residual electric fields such as those present in all laboratory environments and the presence of background ions mix these levels. In this process higher ℓ and m_ℓ states are populated. In the extreme, these high ℓ and m_ℓ states become very long-lived since the electron no longer penetrates the core. We call these high-ℓ, m_ℓ Rydberg states ZEKE states. Hence the transformation of the optically prepared Rydberg states (low ℓ) to ZEKE states occurs only in a small band, typically for $n \geq 100$, within which the density is high enough to be scrambled by the field. It is the detection of this narrow band that is under discussion here.

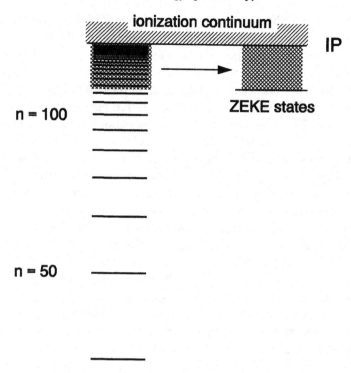

Fig. 6.3 One representative Rydberg series, here leading to the ionization potential (IP). The Rydberg states starting around $n=120$, by interaction with residual fields, are converted to long-lived ZEKE states shown on the right-hand side.

It must be remembered not only that the ground state of the ion has a Rydberg series leading up to it with a narrow set of ZEKE states above $n>100$ in an 8 cm^{-1} bandwidth, but also that this is true for all the individual excited ionic states above the IP. Each one of these has its own Rydberg series leading to a narrow set of states above $n>100$. Each of these bands in turn is extremely long-lived. The new feature is now that this stability is retained even though the total energy is well in excess of that required for

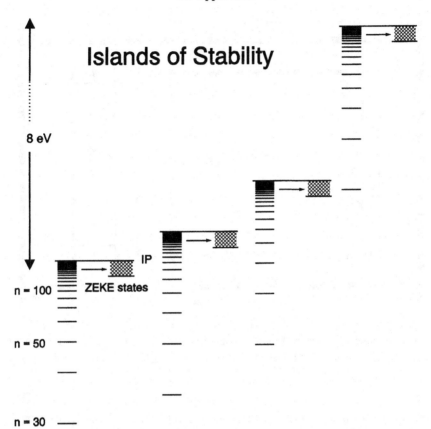

Fig. 6.4 Islands of stability. Here the ionization energy and three additional states at higher energy are shown as an example to show that each state has its own individual Rydberg series. The states at high n, typically near $n=100$ are converted by external fields and ions into special ZEKE states that have an abnormal lifetime, typically some three orders of magnitude longer than one would expect for these states by extrapolation from low n. These abnormally long-lived ZEKE states then exist as special islands of stability at high excess energies within the ionization continuum with a typical bandwidth of some 8 cm^{-1}, as depicted here. These bands can arise from a band of rotations, vibrations or electronic degrees of freedom.

auto-ionization. This is apparently true for states far up into the ionization continuum. These represent 'islands of stability' embedded deep within this continuum. These are states that, contrary to standard wisdom, are long-lived at large excess energies (Fig. 6.4).

Since ZEKE states are usually in a minority compared with directly ionized molecules, the electrons and ions from the latter have to be

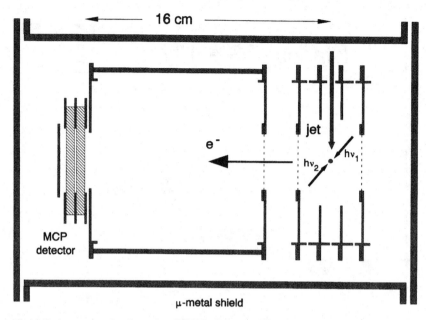

Fig. 6.5 Apparatus for detecting ZEKE electrons. The draw-out pulse is applied as a voltage ramp 1–2 μs after the laser pulse. The electrons drift some 16 cm to a multichannel plate detector in a nearly field-free region (created by use of a μ-metal shield). Two-colour excitation usually simplifies the results.

eliminated to permit measurement of the ZEKE states. Usually the stray fields suffice for this, although a small additional field might be useful.

A typical ZEKE apparatus is displayed in Fig. 6.5. A pulsed field is applied across the ionization region some 1–2 μs after the laser pulse. The delay allows the directly produced electrons to disappear and the pulse ionizes the surviving neutral ZEKE states. The electrons thus produced drift some 16 cm in a nearly field-free region to a multichannel plate detector, where the signal is detected.

The extraction pulse will lower the ionization threshold (in cm^{-1}) by $4\sqrt{F}$ (where F is the field in V cm^{-1}) and thus empties out all the ZEKE states up to this value (Fig. 6.6). These transitions are diabatic in nature.

Hence, for a typical field of 1 V cm^{-1}, this leads to a rather poor resolution of some 4 cm^{-1}. Larger fields may not lead to a much worse resolution since these states with a lower principal quantum number n will be short-lived and hence not survive the pulse delay time.

To improve on this resolution as well as to improve the intensity one can just decrease this ionization pulse height. When this reaches some 50 mV

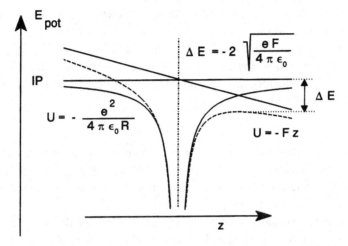

Fig. 6.6 Lowering of the ionization potential due to an electric field.

cm^{-1}, then $E=4\sqrt{F}$ or 0.9 cm^{-1}. This resolution is adequate for many experiments.

6.2 Increasing resolution – slicing

The resolution can now be improved by the method of slicing. In general a simple way to achieve slicing is to apply a slowly rising extraction pulse, preferably with a voltage ramp leading to a slow rise in the field. The early part of this ramp will lead to ZEKE states ionized at lower F arriving early at the detector, the ZEKE states ionized with the larger F corresponding to the late part of the ramp and hence arriving with a delay at the detector. Hence, by setting a certain time gate on the detector, one chooses a certain voltage range ΔF for the extraction pulse.

Hence only the range of F to $F + \Delta F$ will be detected. This in turn corresponds to a narrow range of ZEKE states being detected. This then takes just a slice out of the ZEKE-Rydberg spectrum. This slicing greatly improves the resolution of the ZEKE spectrum, typically down to some 0.2 cm^{-1}. This then is now limited by the typical laser bandwidth of a dye laser and not by the detector. With a programmed pulse generator the above technique can be technically refined further.

In a typical experiment today one looks for the signal between two fields, F_1 and F_2. Since ionization is diabatic, these two fields open up a window

$$\Delta E = 4\sqrt{F_2} - 4\sqrt{F_1}$$

For $F_1 = 50$ mV cm^{-1} and $F_2 = 100$ mV cm^{-1} this would in principle amount to 0.4 cm^{-1} resolution.

It is important to realize, however, that this formula is not correct when F_1 and F_2 are nearly identical. Not only does the intensity disappear, but also, more significantly, due to residual Stark splittings, the limiting resolution in this limit becomes[36]

$$\Delta E = 0.8\sqrt{F}$$

For $F_1 \simeq F_2 = 75$ mV cm^{-1} this becomes 0.2 cm^{-1}. To improve on this value it becomes necessary to reduce the total field. Hence to attain a resolution of 0.1 cm^{-1} one has to keep the fields below 16 mV cm^{-1}. This is possible for electrons but not for ions.

For ions the problem is somewhat more difficult since the direct ions now have to be spoiled by an added fiduciary field. In the process of applying this field, however, one also empties out all high-n ZEKE states, so, as the field increases, as it must for larger masses, the entire band of ZEKE states is spoiled together with the direct ions and nothing is left for detection. A field of 1 V cm^{-1} dips down 4 cm^{-1} into the ZEKE states, making the damage almost complete.

A second technique that can improve resolution is polarity switching. Consider the ZEKE states being split by the Stark field (Fig. 6.7). For a diabatic transition, which is the typical situation for almost all ZEKE spectra at high level densities, one has produced a range of energetic states from red states to blue states. This expectedly smears out the ZEKE spectrum as a result of the Stark field. For the red states, the electron is located to the side of the saddle point of potential and therefore these states ionize more readily than do the blue states, for which the electron sits on the other side. Hence, at a certain voltage the red states will typically ionize in nanoseconds, whereas the blue states will typically ionize in microseconds. The blue states will ionize somewhat faster if an excess voltage is applied. Hence the delay can be shortened by increasing the field above that theoretically required in Fig. 6.7. This means that, when the field is increased substantially, both blue and red states will be ionized.

If one now reverses polarity, the blue states become red states and can in turn be ionized promptly. Hence, one can dump the red states by applying an electric field and then reverse polarity and thus now measure the converted blue states. One thus cuts the spreading due to the Stark field roughly in half. In summary, by applying a pulse to dump the red states and then reversing the polarity of the pulse, one can measure the converted blue states with about half the linewidth.

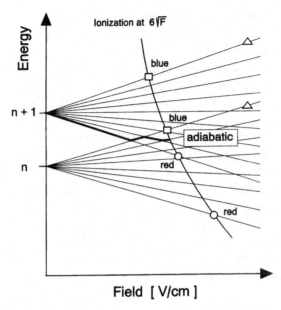

Fig. 6.7 Splitting of Rydberg states as a result of an electric field, showing high-energy blue states (□) and low-energy red states (○) for each Rydberg state in the diabatic limit. Red states ionize as shown, promptly. Blue states are so slow to ionize that they ionize at excess energy (△). The adiabatic limit is nearly at constant energy, ionizing at the line.

Hence improved laser resolution could now more easily recognize individual Rydberg states. On the other hand, a given field again sums all states, the sum of broad states or the sum of narrowed states. This alone will not improve resolution. If, however, the pulse has sufficient excess field strength or exists for some 50 μs then in Fig. 6.7 both blue and red states for the higher ($n+1$) Rydberg state are dumped. Hence this becomes an indirect method of slicing.

If the polarity-reversed field is larger in absolute magnitude than the original field, then this second pulse not only will display the blue states from this original field, but may actually produce new red states in addition. This is again an indirect slicing procedure.

For ions one can employ a bipolar pulse in the mass-detected ZEKE experiment described above, of some 50 mV cm^{-1}. Typically this pulse exists for some 20 μs in order to eliminate all the direct ions. Unfortunately, however, this long a pulse ionizes both red and blue states of the ions below this field, hence polarity switching is not possible here. The next states at lower n, however, survive and can be measured in the same $50 + \Delta F$ mV

cm^{-1} window. If the voltage goes higher, say to $100\ mV\ cm^{-1}$, the resolution decreases to $1.3\ cm^{-1}$, which is the typical problem of loss of resolution for mass-selected ZEKE analysis.

We then have two quite different methods of improving resolution: (a) slicing, i.e. taking a slice of ZEKE-Rydberg states by producing a voltage slice for extraction, and (b) polarity switching, which essentially narrows the range of Stark states detected and thus sharpens the peaks. One can clearly imagine combinations of these two techniques.

Finally, of course, it must be remembered that, when the range produced by this Stark splitting is smaller than the Rydberg level spacings, another alternative is to push the resolution to the sub-Doppler region[37]. Then the states can be identified directly, but each still with a considerable broadening much beyond the width of the laser, this being due to the total range of Stark states populated or the partial range left over by the polarity switching. Here slicing would not be expected to improve matters, but polarity switching should narrow the lines.

6.3 Long ZEKE beam detection

We here want to propose a new technique for measuring ZEKE spectra. This makes use of the discovery that, if treated correctly, the neutral ZEKE states are extremely robust and survive for up to 80–100 μs. These are the original neutral ZEKE states in the absence of any applied field and hence have not yet been separated into electrons and ions by a field. The residual field, whether electrical or magnetic, is harmless since we keep the ZEKE states together throughout the entire flight path. As such they are neutral and not affected by residual fields. They will remain in their original ZEKE state until the end of the long beam. If the ZEKE states hit a positively charged channel plate they will produce a ZEKE electron signal. If they hit a negatively charged channel plate they will produce a ZEKE ion signal. This long-beam experiment is totally ambivalent until the end, when it turns into an electron or ion device, depending on the polarity.

The beam need not be specially shielded against electric or magnetic fields since they do not affect ZEKE states. On the contrary, stray fields are now desirable since they eliminate the unwanted electrons and ions which would obscure the ZEKE signal. Typically we increase the stray field to some $50\ mV\ cm^{-1}$, a value adequate to reject all direct ions in the benzene experiment.

Experimentally one should only note that, if ZEKE states are to survive some 100 μs, they should not collide with walls or apertures, i.e. a carefully

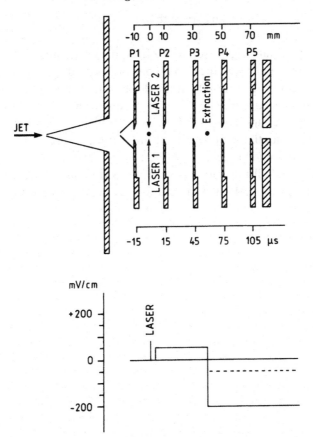

Fig. 6.8 Neutral ZEKE beam apparatus. This is a scaled drawing of the apparatus
in use. The bottom graph shows the pulse sequence for collecting the positive ions
Note that this apparatus should not be shielded.

constructed molecular beam some 20 cm in length is required (Fig. 6.8).
Usually 100 μs suffices to eliminate the ions or electrons that have also been
produced by the laser. Thus this neutral ZEKE beam has, after some 100
μs, been stripped of all direct ions and electrons.

An important experimental aspect of this new technique is the lack of
shielding for electrons or ions. All previous high-resolution techniques
required careful μ-metal shielding. If desired a mass spectrometer, such as
a RETOF, could be connected to this long beam to achieve mass selection
of the ions produced. The long beam in this way could become the front
end of a mass spectrometer to collect all ions or only ZEKE state ions, as
desired, with varying intensity and resolution.

The field at the detector is so chosen that it is slightly above the added 25 mV cm^{-1} field, say 30 mV cm^{-1}. Thus the extra 5 mV cm^{-1} acts as a slicing field required to obtain high-resolution ZEKE data. Technically, the field being on for 100 μs ionizes all the red and blue states produced by the Stark field of 25 mV cm^{-1}. A short field at the detector now cleanly detects the red states in the slice of 5 mV cm^{-1}. Since the question of electron versus mass resolution is left until the end of the ZEKE beam, where now the field imposition and hence the identification of the charged particle occur for the first time, both detection schemes, for electrons or mass analysis, have the same resolution at full intensity, in an unshielded configuration. In fact some stray laboratory field is now a desirable feature of this method.

7

Threshold ion detection

Steradiancy detection for electrons is shown on the left-hand side of Fig. 5.1, together with a selector to discriminate against the straight-through component. This figure was taken directly from Peatman's doctoral thesis[34]. It should be noted, however, that one has

$$A + h\nu \rightarrow A^+ + e^-$$

at threshold. Hence, for every threshold electron there is a corresponding ion at threshold. Since the threshold electron is now a signal that a specific quantum state has been produced, the corresponding ion is similarly in the known state and hence also state-selected. Hence, a coincidence experiment, such as that carried out by Peatman and developed by Baer *et al.*, which measures ions in a time-of-flight mass spectrometer in coincidence with a threshold electron, constitutes a technique for mass-analysed threshold ionization detection. As such it is also a state selector for ions. In the version as a two-particle coincidence technique it becomes then a Photoion-Photoelectron Coincidence measurement (PIPECO)[38]. With the appropriate delay all this can also be readily extended to ZEKE experiments. It should, however, be noted that such coincidence schemes depend on a CW, or near-CW, photon source such as that used by Peatman, or can be extended to synchrotron light sources, as was done by Guyon and Baer. Accordingly, coincidence ZEKE experiments with a CW or synchrotron light source are equally feasible. Initial work on this has been carried out by Hsu *et al.*[39]. As light sources with higher repetition rates (above 1 kHz) become available, coincidence ZEKE will become the method of choice for this purpose.

The ZEKE state is a neutral state that is detected by field ionization. This produces electrons and ions pairwise, either of which can be detected depending on the polarity chosen. The measurement of electrons or ions presumes, however, that there are no electrons or ions produced directly by the laser or from fast auto-ionization, which would interfere with the new electrons or ions produced from the neutral ZEKE states. With the electrons this is no problem because they easily disappear. In general, however, there is a large background of direct ions, which interfere with the positive ZEKE detection of ions unless they are removed from the system prior to ZEKE ionization or separated by virtue of their steradiancy, time delay, etc. Johnson suggested that one way to remove these direct ions in order to avoid this interference would be by the application of a field[40]. For electrons, the use of stray fields or steradiancy detection suffices. For ions, owing to their larger mass, a substantial field is required in this MATI technique. Such a field, however, can create a problem insofar as this field not only will remove the direct species, but also ionizes the ZEKE states and removes them as well, leaving little if anything for final ZEKE detection.

Johnson first demonstrated this with mass measurements on the ZEKE states of pyrazine[40] employing a field of 0.8 V cm^{-1}, which leads to a measurable reduction in signal strength compared with that of electrons. To understand this point one must remember that the ZEKE states exist principally within a very narrow range of some 6 cm^{-1} (half width), just below the IP, or any other state for that matter. This is the great virtue of ZEKE states because as such they are a signature of the IP or the relevant state. There is, however, also a disadvantage to this method. If one were to apply a field for the removal of direct ions of only 2.25 V cm^{-1} one would empty all the ZEKE states within this 6 cm^{-1}, leaving almost nothing for later ZEKE detection, the IP being lowered by $E = 4\sqrt{F}$ (E in cm^{-1}, F in V cm^{-1}).

This field then removes all states from the IP down to $n = 135$ on the Rydberg ladder. Of course, higher fields will still detect lower n, but this is now in the wings of the ZEKE band at greatly reduced intensities. Thus 20 V cm^{-1} has been used[41] to measure the ZEKE masses of nitric oxide, but here the signal is very weak indeed and the spectrum becomes very different and hard to interpret. This is quite understandable insofar as ZEKE measurements should be constrained to be within the normal ZEKE band for optimal intensity and resolution. Perhaps a reasonable upper value for fields that can be safely used without loss in intensity might be 1.5 V cm^{-1}. Even there, all states down to $n = 150$ are lost to detection.

The voltage required to eliminate the interfering direct ions will, of course, increase with the mass of these ions. So, whereas 0.8 V cm^{-1} is used for pyrazine at relative molecular mass 80 daltons, the 1.5 V cm^{-1} limit would be reached near 150 daltons, this effect scaling roughly linearly with the mass. Higher masses can only be measured by going into the wings of the ZEKE band at greatly reduced intensities, as mentioned above. This puts a mass limitation on the method.

The new long ZEKE beam technique constitutes a way around this problem (see above). Here the new fact that the ZEKE states can be left intact and not ionized until the end of the beam, typically after some 50–100 μs, is employed. On this long time scale even stray fields or fields comparable to this, of only 25 mV cm^{-1}, suffice to remove all electrons and all direct ions up to a relative mass of 100 daltons. Hence, again, using the linear scaling law, this limiting mass could now be increased 60-fold, before the attrition of ZEKE states sets in at 1.5 V cm^{-1}. This means that the ZEKE beam technique permits full (highly resolved) ZEKE signals for electrons or ions up to the kilodalton range, employing either electron or ion analysis depending on the polarity at the detector, including full mass resolution as determined by the appended mass spectrometer. The interesting side benefit is that no unusual shielding precautions are required – not even for the electrons. This constitutes a unique and substantial experimental simplification. One might even argue that the long-beam technique is blind with respect to whether an electron or mass ZEKE experiment is done. This decision is delayed until the detector, where the polarity alone decides the process. The apparatus is now ambivalent until the end of the beam.

8

Basic applications

Let us consider the applications. The first test is to measure ionization potentials. That this works is shown in Table 8.1. The customary and perhaps best way to measure accurate ionization potentials is to look at well-defined Rydberg series, because they converge to the ionization potential (IP). The levels get closer and closer and closer, but break off suddenly usually before the ionization potential. Since these energies cannot be resolved at the IP, one must extrapolate. One employs the Rydberg formula to fit the lines and to obtain the IP. Unfortunately, an accurate determination is often much more difficult, since, to obtain this answer accurately, a rotationally resolved Rydberg spectrum in which the ℓ series can be identified is needed. This one rarely has. Specifically, you need the $J=0$ or $N=0$ part of the spectrum. In many cases, obtaining this information has really been the lifetime's work of many people. One of the prominent examples is the case of NO. The task of obtaining these Rydberg series accurately, even for a simple case like nitric oxide, has been the prominent work of Miescher in Basel[42]. In the ZEKE spectrum, on the other hand, one obtains IPs quite directly and with equivalent accuracy without even looking at Rydberg series. The resolution one achieves with ZEKE spectroscopy is such that the values differ from highest accuracy values by two tenths of a wavenumber in the last digit, which is well within the error of any present method. Thus, ZEKE spectroscopy becomes a quick way of obtaining a highly accurate ionization potential directly. It has become not only an easier, but also a superior, way of obtaining such ionization potentials, particularly if no Rydberg levels can be identified (which is typically the case, particularly at high n). However, in some cases one can even reverse the methods and go back to look at Rydberg series based on the final value of the IP from the ZEKE spectrum. This is still a technique of highest accuracy if sufficient resolution is employed. Using sub-Doppler

Table 8.1. *List of molecular ionization potentials determined by ZEKE spectroscopy and other methods*

| Molecule | Ion Energy (cm^{-1}) | | | |
	ZEKE	Reference	Other methods	Reference
NO	74721.7±0.5	35	74721.5±0.5	42
			74721.7±0.5	43
Benzene	74555.0±0.4	44	74573.0±2.0	45
	74556.1±0.3	46	74528.0±15	47
Benzene–Ar	74387.3±2	48	74388	49
pDFB	73870.4±2	50	73871±5	51
pDFB–Ar	73636.5±2	52		
Phenol	68628.0±4	50		
Phenol–H$_2$O	64024.7±4	53	64035±10	54
NH$_3$ (v^+=1)	83062.2±2	55	83305	56
NH$_3$ (v^+=2)	84000.8±1	55	83897	56
O$_2$	973348±2	57	97358	58
Na$_2$	39477.9±1	59	39478.7±0.1	60

Note:
pDFB, p-difluorobenzene

excitation Neusser *et al.* were able to identify 69 Rydberg series for benzene[61].

Now, in a certain sense, this indicates that ZEKE methods simplify the determination of IPs while maintaining very high accuracy. If one looks at these high Rydberg states carefully, the experiment is a little bit more complicated than one might think at first, which at the start of this work caused many problems. In the meantime, it has become a great blessing. Let us look at Fig. 8.1. Above the IP, molecules ionize. Below the IP, a molecule is in a high Rydberg state at n=100–300. It is readily ionized with a small pulsed delayed electric field. At lower n, one has the normal Rydberg region. In this region the lifetime of the Rydberg state is dominated just by the orbiting frequency sweeping by the core. This goes as n^3 in this low-n region. When one looks at lifetimes of the Rydberg states near the ionization potential, it is well known that these lifetimes become longer as n^3, where n is the Rydberg index. The subtle new point is that the ZEKE state lifetime cannot be predicted by this formula; it typically deviates from this value by some two orders of magnitude. If one extrapolates lifetimes to high n, say n=200, these lifetimes extrapolating as n^3 should be in the range of maybe 200–500 ns, which just is not so. When we measured some of these ZEKE lifetimes, which was done recently, we came up with numbers like 20–50 μs.

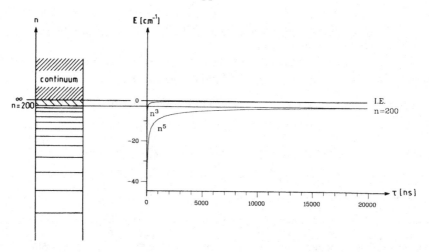

Fig. 8.1 Regions of ionization, showing the special high-*n* region used for ZEKE spectroscopy and the lower *n* states, together with their lifetimes. In the transition region to the ZEKE states there can be a further lengthening of lifetimes. They can change abruptly by up to two orders of magnitude.

If one waits 100 µs, one can still see these states. The interesting fact, at least for a molecular system, is that the high-*n* states all survive anomalously long, which is a property characteristic of the last Rydberg states 'hugging' the ionization potential from below. This was, of course, the reason for the success of the new ZEKE beam technique mentioned above. This was first seen in our laboratory by G. Reiser *et al.*[35], who discovered this most interesting new effect of the extreme longevity of all ZEKE states. These states are extremely long-lived and if one does not take cognizance of that, then the IPs are wrong by some 10 cm^{-1} because one has not waited long enough.

So, lifetimes in the ZEKE region experience an additional lengthening due to the strong coupling among these high-*n* states which more reasonably predicts an n^5 dependence on lifetimes under this regime. This can be nicely seen by looking at the energy dependences as they are shown in Fig. 8.1 and has been explained in detail by Chupka[62].

The interesting and unexpected result now is that, whereas the low-*n* states go as n^3, this switches suddenly to the n^5 behaviour the moment the coupling of the levels sets in. Hence the direct manifestation of ZEKE states is a break in the lifetimes curve as they suddenly jump from 20 ns to 20 µs, a jump of over three orders of magnitude. This sudden jump is one of the clearest and most unexpected results in ZEKE spectroscopy and is nicely observed in the DABCO (diazabicyclooctane) spectrum (Fig. 8.2).

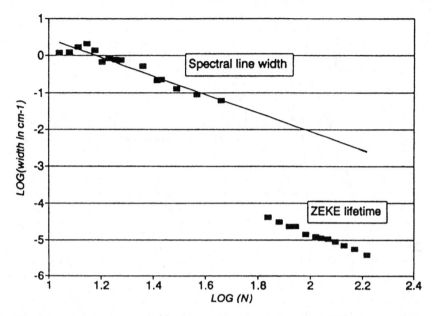

Fig. 8.2 A log–log (base 10) plot of the Rydberg states of DABCO versus n. This shows the dramatic jump undergone by the lifetimes as they reach the ZEKE region[63]. The low-n values follow the well-documented n^3 law as observed for benzene[37], even though here line width data are shown.

Once one realizes this enormous lifetime enhancement one can turn the whole question around and say, since one knows this, one just needs to wait for 10 or 100 μs, until electrons and ions have gone away as charged particles and then look at the last long-lived Rydberg states 'hugging' the respective ion states from below. Each ion state will thus be 'hugged' from below by these special Rydberg states which we term ZEKE states. The last Rydberg states can be formed quite close to the true ionization potential. This is the easiest way to measure the spectrum of ion states because one is looking at the very last, high-energy Rydberg states which are then finally ionized by the field. As a practical point one must note that measurements extremely close to the IP become increasingly more difficult in the limit, due to stray fields. The practical answer to this is to calibrate the system for varying small fields and to measure near $n=200$.

In this particular experiment, ZEKE spectroscopy and total current measurement again are really quite different. One can say that the total current signal must also include the ZEKE signal, but without the filtering process of the ZEKE technique these peaks are buried in the signal. Let me give you the simplest of all examples you can imagine, the ionization of

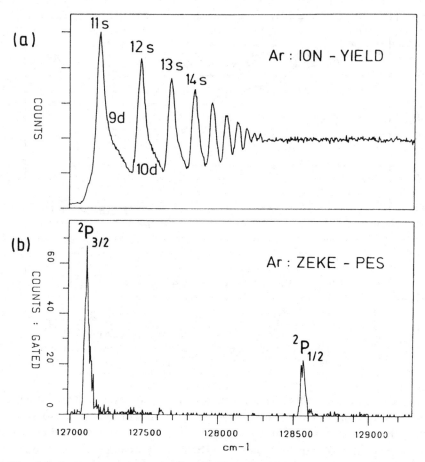

Fig. 8.3 Argon atom ionization[64]. (a) The photo-ionization efficiency curve, showing Rydberg structure leading up to the (not seen) $^2P_{1/2}$ state. (b) The threshold ZEKE spectrum with the Rydberg series suppressed, but clearly showing both $^2P_{3/2}$ and $^2P_{1/2}$ thresholds. Note that the $^2P_{3/2}$ ionization threshold is hidden below the 11s Rydberg peak in the PIE spectrum.

atomic argon (Fig. 8.3). One pumps in more and more energy optically until argon ionizes at the $^2P_{3/2}$ state of the IP, thereby generating a current as shown in Fig. 8.3(a). After this, the signal goes up and into an auto-ionizing Rydberg series. The Rydberg series collapses at higher energies. This converges to the second state of the argon ion, the $^2P_{1/2}$ state. The start of the entire spectrum and the onset of the current arise from the $^2P_{3/2}$ state; subsequently, one sees the Rydberg series with the electrons converging to the $^2P_{1/2}$ state. One will not know where the $^2P_{1/2}$ is unless one can perform

a good Rydberg extrapolation. For obtaining the ZEKE spectrum in Fig.
8.3(b) one simply imposes a delayed field and switches on the ZEKE detector to measure to see how it extracts this totally hidden level from this particular spectrum. One now sees the peaks due to the $^2P_{1/2}$ and $^2P_{3/2}$ states
and very little else. The Rydberg series in the electron spectrum again cor-
relates to hot electrons and they are suppressed by the ZEKE technique.
One sees that not even this simple spectrum can be analysed since there is
not even a staircase function to extract these two peaks from the spectrum.
This example demonstrates the effectiveness of the ZEKE filter even for the
very simple case of an atomic spectrum. The interest here is really derived
from molecules. As a simple case one can choose a diatomic molecule, nitric
oxide in Fig. 8.4. This brings out another new feature. First of all, one sees
in Fig. 8.4(a) Turner's photoelectron spectrum[65] of nitric oxide with the 21
eV He line. This shows the first five vibrations of the ground state in the
nitric oxide ion at the medium resolution of early experiments. One can
compare this with the spectrum obtained by measuring the total current, as
Watanabe did (Fig. 8.4(b)). One can see the onset of the first staircase to
the $v^+=1$ and $v^+=2$ vibrational states and then it smooths out. Later there
is a complex mixture of auto-ionization and Rydberg structure, which
really makes this a most difficult spectrum to analyse. There is an extensive
literature analysing photoelectron spectra with high-resolution photon
sources to uncover underlying states and Rydberg structures[66].

One now turns on the ZEKE spectrometer to obtain the spectrum shown
in Fig. 8.4(c). First of all, one sees the same five lines as in the photoelectron spectrum. Now a new effect occurs. The ZEKE spectrum persists up
to $v^+=26$. One has unexpectedly pulled out the ZEKE spectrum from the
sea of electrons which produced the congestion of the PIE spectrum at
higher energies. However, interestingly, as a totally new effect one can see
parts of the vibrational ion spectrum which one does not see in photoelectron spectroscopy. The intensity of the vibrational transitions in the ion is
quite surprising since they are not predicted by the Franck–Condon principle. This is referred to as the so-called Franck–Condon gap. A similar
situation has recently been discovered for nitrogen, whereby vibrational
levels up to $v^+=67$ are observed within the Franck–Condon gap[67]. Two
interesting questions are the following: how does one enter this
Franck–Condon gap and how does one obtain this particular part of the
spectrum? The answer must lie in channel interactions that permit one to
gain oscillator strength in the ZEKE spectrum that does not exist in any
normal spectroscopy. Hence a ZEKE spectrum does not just produce a
better version of a photoelectron spectrum. It gives absolute accuracy

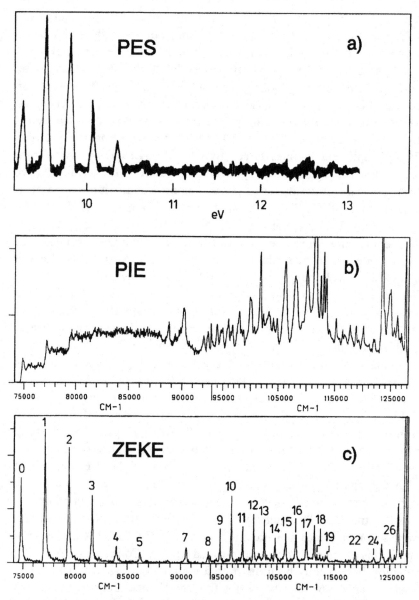

Fig. 8.4 A comparison of methods for nitric oxide. (a) The photoelectron spectrum from Turner, (b) the staircase function of PIE and (c) the threshold-ZEKE spectrum. Note that the vibrations go up to $v^+=26$, whereas only $v^+\leq4$ is seen in the PES. This demonstrates a fundamental difference between the two spectroscopies.

derived directly from the photon source in laser spectroscopy, but it in addition, as a new feature, reveals transitions that are otherwise not observable, which is a very important point. In addition, as one will see below, it offers another benefit – some 1000-fold better resolution.

On going to high resolution one can make yet another discovery in ZEKE spectroscopy: one can now expand the $v^+=0$ peak, which is the ionization onset for the nitric oxide ion. Here one observes an unexpected very large increase in resolution (Fig. 8.5) and observes that this single vibrational peak is in fact split into three peaks (our early 1984 spectrum[68,69] is represented in the lower trace). When one puts Turner's spectrum on the same scale for $v^+=0$, one obtains the top spectrum. In our high-resolution spectrum, we show only the range of some 3 meV, which is the width of the line drawn above in the middle spectrum. The width of this marker-pen line is what is amplified here. This is the primordial ZEKE spectrum in which one can see, for the very first time, the onset of the ionic rotations starting from the origin $N^+=0$, 1 and 2. This was the first rotationally resolved spectrum at the ionic origin and thus represented a landmark in electron spectroscopy. The resolution is an improvement by some three orders of magnitude.

The question that immediately puzzled us was that of why there were three peaks. Why does one see these high-angular-momentum states and why does one not just see the nearest angular momentum components for the outgoing electron? Rotational excitation has intensities due to partial wave contributions to the outgoing electron $|k\ell\rangle$, for which conservation rules for angular momentum[70] state that

$$\Delta N = N^+ - J' = \ell + \tfrac{3}{2}, \ell + \tfrac{1}{2}, ..., -\ell - \tfrac{3}{2}$$

where $\tfrac{3}{2}$ is the sum of the spin and photon contributions to the angular momentum. Hence, higher angular momentum components just reflect the higher ℓ states that are populated. In more refined work, we obtained more data that are shown on the left-hand side of Fig. 8.6. Here we start from various rotational states at the S_1 origin: starting from $J'=0$, the $N^+=0$ transition is prominent; starting from $J'=1$, the $N^+=1$ transition is prominent and so on. However, in addition, there is an intensity alternation that is simply the parity alternation. This is given by the parity selection rules for rotational ionization[73]:

$$\Delta N = \Delta p + \ell = \text{odd}$$

here given for Hund's case (b) in which Δp is the change in electronic parity upon ionization. How can one understand the intensities of these

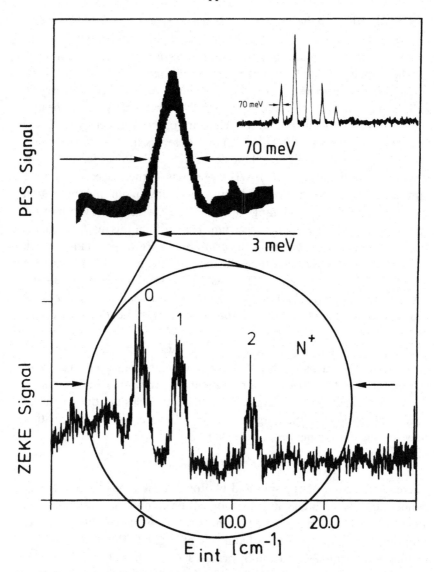

Fig. 8.5 The primordial 1984 ZEKE spectrum of nitric oxide. Note that the thickness of the marker-pen line is expanded into the ZEKE spectrum below, clearly showing the initial rotations in the ions for the first time. Note the high-angular-momentum components. The inset is from the book on the $v^+ \geq 0$ states by Turner et al.[65].

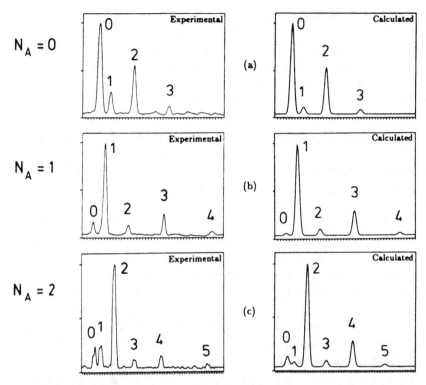

$N_A = 0$ (a)

$N_A = 1$ (b)

$N_A = 2$ (c)

Fig. 8.6 The ZEKE spectrum of nitric oxide from defined initial rotation states in S_1. The plots on the left-hand side are from experiments[71]. The plots on the right-hand side show the results of calculations by Rudolph *et al.*[72] using HF centrosymmetric wavefunctions without electron correlation. This was the first correspondence between ZEKE experiments and theory, showing the alternation of the higher angular momentum components for various initial vibrations.

high-angular-momentum states? Some preliminary calculations confirmed this result[74], but the detailed work of Rudolph, Dixit and McKoy at Caltech (shown on the right-hand side of Fig. 8.6) then provided *ab initio* calculations using Hartree–Fock centrosymmetric wavefunctions even without electron correlation that explained these alternating intensity data[72]. Multichannel quantum defect theory (MQDT) was already successful at just matching our spectra[74]. This at least convinced us that what we are seeing in this new spectroscopy is real and not an artefact. In more recent work, Akulin *et al.* from Moscow suggested a direct physical model[75] for this process. The ZEKE spectrum was made plausible by a simple orbital hopping model with scattering at the ion core. This directly predicts the experimental spectrum. After photon absorption the molecule is described in terms of the motion of a free (or a Rydberg) electron in a Coulomb potential and the motion of the molecular ion core. The interaction between these motions is described by the scattering of the electron at the ionic core which can take place when the electron approaches the ion. It results in non-adiabatic horizontal transitions, in which energy and angular momentum are transferred from the electron to the ionic core and vice versa. This reproduces the experimental spectrum very accurately with just two reasonable parameters: the ionic charge asymmetry and the inter-atomic distance within the nitric oxide molecule. The latter is known spectroscopically, and the former was confirmed in a calculation by Rosmus[76]. The two externally known parameters produced surprising agreement with the observed spectrum, shown in Fig. 8.7.

In this case, theory and experiment come together. In fact, one could say that the role of theory has become of essential importance to us in interpreting these new kinds of spectra. Without *ab initio* calculations, it is impossible for us to assign our new ZEKE spectra accurately. Prior computational techniques were developed using the transferability of force constants as in sophisticated molecular modelling and available on workstation packages. They did not work for these spectra. The predictions from these programs are hundreds of wavenumbers off from the observed spectra. As a consequence nothing can be learned from these modelling calculations for these ZEKE spectra, although they are useful for the interpretation of normal spectra. One really must do sophisticated *ab initio* calculations although correlation may sometimes be neglected and then the Hartree–Fock method is adequate. This gets close enough in many cases to predict and, hence, to assign the observed ZEKE spectra, particularly for soft modes.

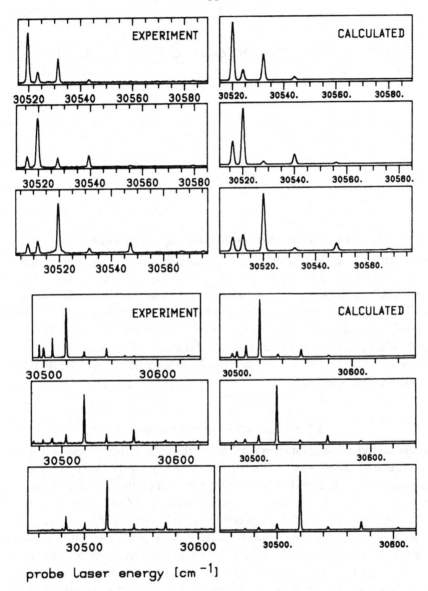

probe Laser energy [cm^{-1}]

Fig. 8.7 Model calculations of the ZEKE spectrum of nitric oxide using a simple orbital hopping mechanism[75].

9

Specific examples of ZEKE spectra in brief

9.1 p-Difluorobenzene – vibrations and propensities

As a practical example, let us illustrate what happens in more complicated systems, as is the case for *p*-difluorobenzene in Fig. 9.1. The top spectrum shows state-of-the-art photoelectron spectroscopic results from Sekreta *et al.*[31] with a very good low overshoot in the energy. Below that, you see the ZEKE spectrum and here one can begin to assign the vibrational bands with some precision in a straightforward manner. Here one can use the Franck–Condon principle to project our knowledge of the S_1 state into the assignments in the ion.

9.2 Nuclear spin isomers in ammonia

If one considers NH_3, for example, one can distinguish the two species very nicely due to nuclear spin. NH_3 is similar to o-H_2 and p-H_2, but in this case there are three identical hydrogens with two permutations. Hence one has ortho–para isomers due to nuclear spin symmetry. ZEKE spectroscopy allows one really to see the two-spectra (Fig. 9.2) from selective intermediate state excitation. The case of ortho-NH_3 is quite distinct from that of the para form as a result of the concomitant rotational levels dictated by the symmetry of the total wavefunction. Naturally, this is useful, because, with this kind of resolution, one can assign the origin of the ionization spectrum of NH_3 very accurately. This has been a matter of some contention[78], which can be readily resolved completely with the *N*- and *K*-type structure that one sees here. This became a text book example for ZEKE resolution of nuclear spin isomers in a mixture. Such measurements can only be performed in mixtures since such nuclear spin species, up to now, cannot be physically separated except for hydrogen.

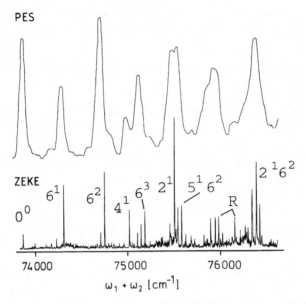

Fig. 9.1 *p*-Difluorobenzene. A comparison of the ZEKE spectrum with the high-resolution PES spectrum with excess energies via the 6^1 vibration of the S_1 state[31,77].

9.3 Benzene – the Jahn–Teller effect

A particularly interesting example is the case of the benzene ion in the $^2E_{1g}$ state, in which the Jahn–Teller effect should show up strongly, but had not been seen. This was first seen in the mass spectrum[79] but recently we obtained the high-resolution ZEKE spectrum, in which we see the highly resolved Jahn–Teller effect (Fig. 9.3). If you look at the photoelectron spectrum of benzene in the top trace, you see just the vibronic progression. In the high-resolution ZEKE, we even see the splitting of the lower half of the 6^1 pair of linear dynamical Jahn–Teller-split states, which is a result of the quadratic Jahn–Teller effect.

9.4 Phenol–water clusters – hydrogen bonding

Clusters are of particular interest. Here phenol and water are combined. Phenol and water stick together with a weak bond. Figure 9.4(a) shows work done by Lipert and Colson[54]. They were looking at the total ion signal (PIE). The staircase function in this case can be improved by using low-field measurements and so we repeated the staircase experiment (Fig. 9.4(b)). This spectrum remains hard to interpret. The resolution can be much

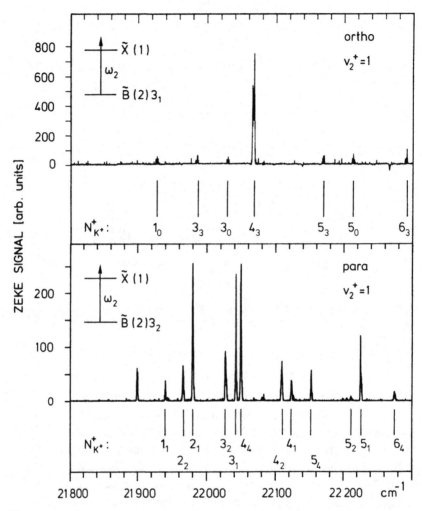

Fig. 9.2 Nuclear spin symmetry demonstrated for NH_3, showing the differing ortho–para states with rotational resolution[55].

improved with a ZEKE experiment and the corresponding spectrum can be seen below in Fig. 9.4(c). We can readily assign the observed lines. Basically, this ZEKE signal is always inherently contained in the total signal already. The spectrum in Fig. 9.4(b) includes all electrons present, whereas all we have done in the ZEKE experiment is put an additional filter into the system to obtain the spectrum in Fig. 9.4(c). It is the same signal that had been buried in the total of all electrons that can just be separated out and otherwise could not have been seen. If one does this even more carefully,

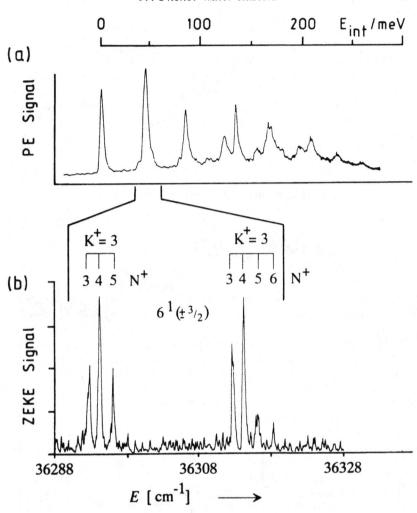

Fig. 9.3 Jahn–Teller splitting in the benzene ion (doubly degenerate) observed (b) and assigned by ZEKE spectroscopy (c)[80]. In (a) is the photoelectron spectrum showing the unsplit ν_6^+ vibration[81].

one can actually use this to assign the new van der Waals modes. As an example, this is shown for a combination of phenol with methanol instead of water (Fig. 9.5). In the spectrum, one can see a great many overtones and combination bands involving the various van der Waals modes in the intermediate states. We have arbitrarily named these modes σ modes or η modes, there being six of these various intermolecular modes involved. Their motions are defined in Fig. 9.6. Some have A′, while some have A″

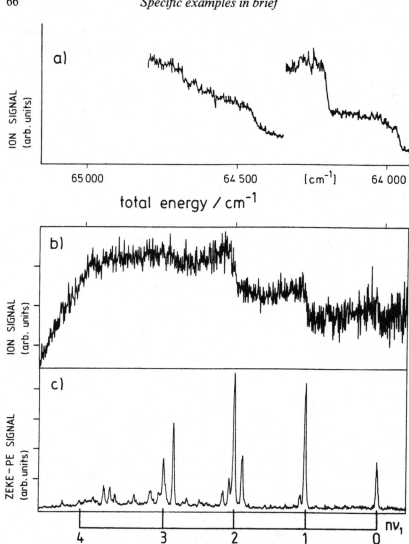

Fig.9.4 Phenol–water cluster spectra: (a) and (b) are PIE spectra; (c) shows the ZEKE spectrum for this hydrogen-bonded complex.

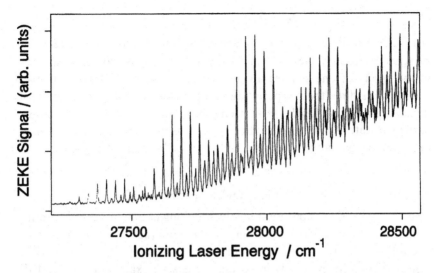

Fig. 9.5 The phenol–methanol ZEKE spectrum of the S_1 0^0 intermediate state, showing a series of overtones and combinations based on just the low-frequency intermolecular vibrations.

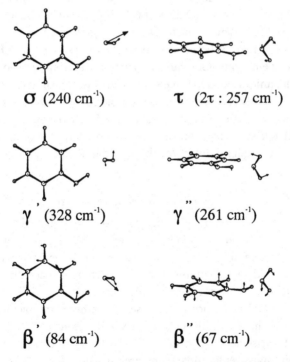

σ (240 cm^{-1})

τ (2τ : 257 cm^{-1})

γ' (328 cm^{-1})

γ'' (261 cm^{-1})

β' (84 cm^{-1})

β'' (67 cm^{-1})

Fig. 9.6 Normal modes for the six intermolecular vibrations between phenol and water and the assignments. Note that all six are observed, even though one could expect selection-rule difficulties in a normal spectrum.

symmetry, the latter being forbidden in direct selection rules. Here, in ZEKE spectroscopy they are apparently not forbidden and we see all six. In the spectrum one can see all the overtones and combinations in the various transitions, thereby permuting the possible overtones and combinations is a useful tool for the assignment of the van der Waals modes in this particular state. Again this visibility of all transitions is without doubt due to the unique feature of channel coupling in ZEKE spectroscopy.

9.5 Free-radical spectra

Another interesting application involves looking at mass-resolved spectra of unusual systems, which are fragile in the sense that they cannot be put into a bottle or an infrared spectrometer for measurement, but are nevertheless chemically extremely important. One such example is the methyl radical cation and Fig. 9.7(a) shows a ZEKE spectrum recorded by Chen at Harvard and White at Brookhaven[82]. In Fig. 9.7(b), the corresponding simulation is shown. You can see that even these kinds of fragile systems can be very readily attacked with ZEKE spectroscopy. This is particularly interesting because the dimension of mass spectrometry can now also be applied to such complicated systems. The area of free radicals in ZEKE spectroscopy is just emerging. One can assign complicated systems for which the traditional techniques that have been used will fail. The importance of such ZEKE spectra has been demonstrated amply for small systems[82–87]. Their measurements can, of course, be mass-selected with the aid of a mass spectrometer; ZEKE spectra of radical systems much more complicated than those shown here will no doubt be discovered in the near future.

9.6 ZEKE spectroscopy of anions and mass-selected neutral species

Anions can be produced by attaching electrons to a neutral species in the expansion zone of a jet, or even by transfer of Rydberg electrons as a particularly soft technique. These anions then can be excited with a laser, leading to photodetachment and a photodetachment spectrum[88,89]. This is analogous to the photo-ionization efficiency spectrum (PIE) of cations, except that the onsets are not a staircase, but rather rise typically as $E^{\ell+1/2}$, which for complicated molecules usually reduces to a $\sqrt{E}$ dependence. Hence the staircase becomes 'rounded'. The ZEKE spectrum of the anion gives us again sharp peaks and as such precise values for the vibrational frequencies of the anion, but also of the ground state of the mass-selected neutral species (Fig. 9.8).

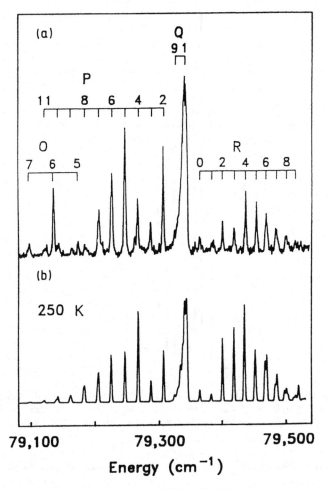

Fig. 9.7 The ZEKE spectrum of the methyl radical ion: (a) experiment and (b) simulation[82].

In this simple way, one can have mass-selected the species via their original anions and then prepared neutral species, which are thus mass-selected. Transitions to selected ground state levels of the neutral species are shown (Fig. 9.8), both starting from vibronic levels of the anion, and to vibrationally excited states of the neutral species at shorter wavelengths. Hence, both neutral and anion spectra are measured in a mass-selected manner.

Mass selectivity is now ensured by the anion leading to the originating ground state. The lower transitions shown in Fig. 9.8 reveal the vibrational

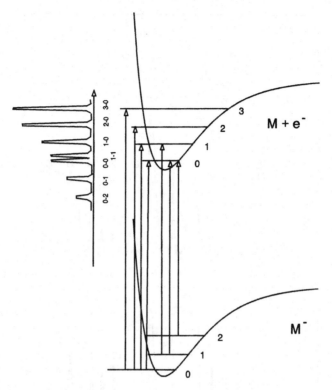

Fig. 9.8 Mass-selective ZEKE spectroscopy of ground state neutral species. Note that the spectra both of M and of M⁻ are obtained. This is mass selection of the neutral ground state, together with its ground state spectrum.

levels of the anion which are detected here with analogous ZEKE spectroscopy.

In addition, one has to remember that neutral species obtained from electron detachment from their parent anions can be very unstable species. These still yield a ZEKE spectrum. Hence, this method is suitable not only for unusual kinds of neutral species that are mass-selected to form a mixture but also for species that may be very unstable. The instability of the final neutral species could result from unimolecular decay, or from the reactivity of the species under study. In general any peak and associated species produced in an anion mass spectrum can be used as a precursor to produce a spectrum of the corresponding neutral species and thus provide for a method of studying the kinetics of such systems.

Let us give a very simple spectrum. Suppose that one measures the anion ZEKE spectrum of OH⁻ (Fig. 9.9), which is a very unstable radical in its

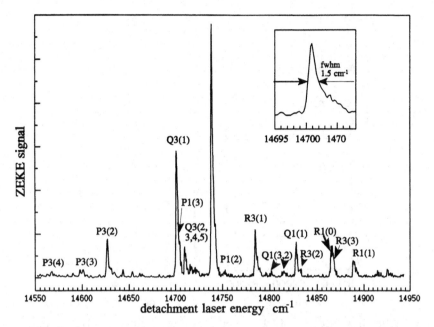

Fig. 9.9 The rotationally resolved anion-ZEKE spectrum of OH^{-90}.

neutral state. In the ZEKE spectrum we now have seen many highly resolved sharp lines, which can be readily assigned.

9.7 Metal clusters

With metal clusters one has some complications. A good example is the ZEKE spectrum of Ag_2^+. One can measure a nice series of vibrational states (Fig. 9.10). The different spectra are obtained via a series of resonantly excited intermediate states. Notice that $v^+=0$ is always the strongest in intensity. If one does a Franck–Condon factor calculation using the best known anharmonicities and atomic distances, the spectrum looks quite different (Fig. 9.10(b)). The calculated peak intensities are way off from the observed ZEKE spectrum. This is a case in which one must have recourse to a model involving channel coupling. In this model, one excites to Rydberg series of other transitions and enters the final states laterally via channel coupling. This is another example of modifying the traditional Franck–Condon process.

This first survey was intended to provide a once-over, thumb-nail sketch of the range of applications in ZEKE spectroscopy. I will go into detail in

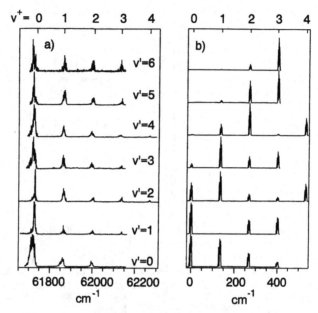

Fig. 9.10 Mass-selected ZEKE spectra of silver dimer: (a) measured vibrational intensities and (b) calculated Franck–Condon factors.

a later chapter. I have tried to show a series of applications of systems that I think are the types of systems that will be of interest in the future: specifically, those systems which cannot be put into bottles, the systems which are fragile, are fluctional, are clusters and the like. I think that ZEKE spectroscopy has become an interesting entrée to the study of systems of this type.

This was to demonstrate that, in the 10 years since the first ZEKE experiment in 1984, ZEKE spectroscopy has become a mature technique but many interesting experiments have still to be done. This opens up a new stretch on our long road to the understanding of new chemical systems. It gives us a new view of the architecture, which is appearing along our road through chemistry, for new species of great interest in chemistry.

The ZEKE method has now found wide application in many laboratories worldwide where many new systems are routinely being found. We are probably only standing at the beginning of many applications of ZEKE spectroscopy. ZEKE spectroscopy should particularly lend itself to studies of systems that do not come in a bottle (i.e. are not blessed with great stability), which is important because the characterization of such chemical intermediates remains crucial to our understanding of chemistry and chemical processes, in particular those involving reactive intermediates.

Lord Todd's comments concerning the changes in the type of chemical systems which will concern us in the next century I think are very prescient and will call for many new ways of investigating chemical systems. Biochemistry certainly has taught us the importance of semi-weak bonds. Thus we require new methods for investigating new chemical intermediates and fragile species. Much needs to be done; the topics discussed so far are only a brief outline of the field to date.

Part II

Mechanisms and applications

10

Historical perspectives and principles

This chapter is intended to give a historical perspective together with certain basic precepts of the ZEKE method. At this juncture in time, ZEKE spectroscopy provides accurate, very-high-resolution spectra of cations and anions, and via ZEKE electron detachment, neutral species. The present resolution of ZEKE spectroscopy is limited only by the bandwidth of the dye lasers employed, e.g. about 0.2 cm^{-1} for cations and about 1.0 cm^{-1} for anions.

The rate of progress has been extremely great in the last 5–10 years as the method has been refined and has become better understood. It has branched somewhat and is now rapidly becoming an interesting new branch of spectroscopy for which many new applications to chemical systems have been developed. Independently, the physics focusing on elucidating the basic mechanism of the ZEKE process itself has appeared and constitutes a rich field in its own right[3,91].

Historically, of course, we started with threshold spectroscopy in 1969, but the breakthrough came with the delayed field extraction experiments which afforded the enormous increase in spectral resolution and resulted in the many subsequent discoveries which provide the understanding of the new method itself. These advances, taken together, form the basis for a new form of spectroscopy.

This scenario resulted from the usual mixture of planning and serendipity leading one to characterize these ZEKE states. Foremost was the discovery of the effect by Reiser *et al.*[35] revealing that anomalously long lifetimes are produced from Rydberg states above $n \approx 100$, which represents a very narrow band of about 1.3 cm^{-1} below the IP.

It now must be recalled that such Rydberg series exist not only up to the IP of a molecule, but in fact for every state in the ionization continuum. So $v^+ = 1$, being the first vibrational state of the ion, lies at an energy above the

77

IP. It will in turn have a complete and independent Rydberg series at energies below. This is so for every one of the very many states of a molecular ion.

We now know that ZEKE states can be produced only from the last few Rydberg states below any ionic state, even for ionic states lying far above the IP that are well within the ionization continuum. These states are long-lived not only below the IP but also for these ionic states high within the ionization continuum. This came as quite a surprise. Again they require a small stray field for their production, as one now thinks. At the beginning this long lifetime was the bane of our existence. More recently, it has become the method of choice, allowing us to measure these highly resolved ZEKE spectra.

The historical background of photoelectron experiments is the history of Photoelectron Spectroscopy (PES) which was pioneered in the experiments of Al-Joboury and Turner at Oxford[92,93], of Terenin, Popov, Kurbatov and Vilesov in Leningrad[94,95] and of Siegbahn in Uppsala[96]; the former principally in the VUV, the latter mainly in the XUV. These pioneers showed what great importance lay in the spectroscopy of ions. Our own initial work in the mid-1960s arose from a desire to liberate photoelectron spectroscopy from electron monochromators and to remove the fundamental limitations found in accuracy, drift and resolving power. In so doing, we came onto a totally new track, which turns out to have many features complementary to photoelectron spectroscopy, including the now obvious increases in resolution and accuracy, attributes that are clearly desirable in any spectroscopy. Independently, we also revealed a host of new and different spectroscopic features, affording a totally new spectroscopic technique.

I was fortunate to have two very able graduate students in Evanston, Bill Peatman and Tom Borne, as well as two distinguished post-doctoral associates, Tom Baer (now at North Carolina) and Paul Marie Guyon (now in Paris). In this work, steradiancy analysis was discovered, albeit indirectly. Although at first the experiment did not work as had originally been planned, the consequences turned out to be far more interesting after we had understood the steradiancy effect. This led us to tune the radiation source through the transition employing this effect. Although the accuracy of this experimental approach was good compared with that of PES, threshold spectroscopy only marginally improved the spectral resolution. It was in the 1984 paper by Müller-Dethlefs *et al.*[68] that we first reported the use of ionization by delayed pulse extraction and found that this opened a new world for molecular ions. The rest is history. Probably the most interesting cases still lie ahead of us but much more sophisticated apparatuses

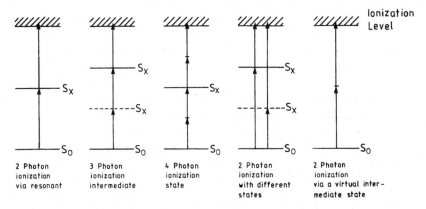

Fig. 10.1 Mechanisms of ionization with two or more photons.

(i.e. new lasers, etc.) will be required. Such laser systems, which are just appearing on the horizon, have tunability into the mid-IR, high-frequency operation, coherent excitation, femtosecond ZEKE analysis capability and much more. Such IR methods with the new OPOs will be of particular interest for anion ZEKE Spectroscopy.

To start, let us review the various methods of ionization. Ionization can be done at various kinds of levels, as shown in Fig. 10.1. One can ionize with two photons via a resonant intermediate state, which is referred to as the REMPI technique. One can utilize 2+1 excitation, 2+2 excitation or various other intermediate states (indeed a virtual state can be used) in order to reach the ionization continuum energetically. The latter is somewhat dangerous since in this case, to obtain adequate signal strength, laser excitation intensities often are sufficiently high that they induce multi-quantum effects. Classically, benzene can be carbonized in this way. Spectra of background gases often obscure the spectra of interest. The resonant multiphoton technique is particularly useful here, for a variety of good reasons. For example, even in a jet-cooled experiment, it is very difficult to select a pure ground state optically unless, of course, it is possible to do some optical labelling experiments. For the case of benzene, the ground state, even in a cooled jet, has the 6_1 vibrational state excited in the electronic ground state. It becomes populated, which means that there may be a 100 K jet expansion. It is very difficult to cool the jet vibrationally, although rotational cooling is readily accomplished. Perfect vibrational cooling is almost impossible. On the other hand, if one can pump to the S_1 state with reasonable selectivity, then a reasonably pure state preparation is achieved here. Thus a clean origin for the second step into the ion can be

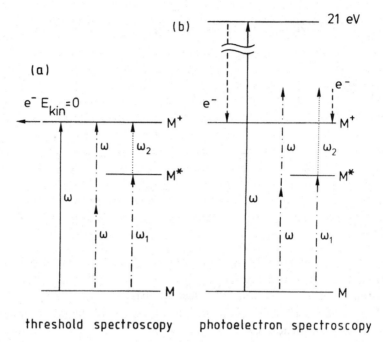

threshold spectroscopy   photoelectron spectroscopy

Fig. 10.2 A diagram showing the difference in excitation mechanism between (a) ZEKE spectroscopy and (b) photoelectron spectroscopy. The latter is an overshoot technique at fixed photon energy analysing the energies of the emitted electrons. The former detects zero-kinetic-energy electrons at threshold only and scans the light source.

obtained. This is helpful since one would like to know from which state one originates in order to know the excitation selection rules. One wants to use this intermediate state not so much because it cuts the wavelength in half, which is a useful but incidental point, but rather because it is really a state-selective step whereby one is able to do a fairly reasonable job of exciting one individual level. This is an important point. In addition state selection allows one to 'walk around' the Franck–Condon energy surface. The three main techniques which I am contrasting then are (i) the photoelectron spectroscopy technique (Fig. 10.2(b)) or over-shoot technique that is due to Turner, Siegbahn and Terenin (PES), (ii) the photo-ionization efficiency (PIE) technique due to Watanabe and (iii) the method of simply scanning the current with the wavelength (Fig. 10.3(a)), using a variable frequency that passes over the threshold while measuring the total current. Our original technique (Fig. 10.2(a)), which is a threshold technique, measures the threshold current and is illustrated by comparison with PIE in Fig. 10.3.

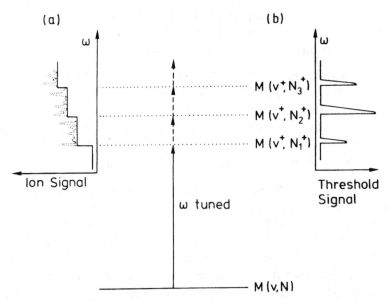

Fig. 10.3 Measurements of photoelectrons scanning the photon energy. (a) PIE, which measures the total photoelectron current. (b) ZEKE, which only measures the zero-kinetic-energy component of the photoelectron current.

Lastly, the several variants of ZEKE in which a delay of a microsecond is introduced between pulsed laser illumination and the drawing out of the electron are considered. If one looks at this question rather more carefully, there are really four variants that need to be considered. Three of these four variants are as follows: (i) the signal at threshold is measured, but only after the amplification of the arrival time differences as a result of the electron being drawn out by the delayed pulse signal (Fig. 10.2(a)); (ii) the somewhat less simple situation in which the state just below threshold (i.e. the last Rydberg state) is examined; and (iii) the positive partner of the charged pair is measured. One generates an electron and an ion in a pair and hence can measure either charge. This last approach has certain advantages and certain disadvantages. For completeness, a fourth technique may be envisioned, but to the best of my knowledge it has never been reported. It requires CW sources and vacuum UV radiation, a set of requirements difficult to fulfil with lasers since it requires high-repetition-rate sources (typically at 10 kHz) or a CW source. This coincidence experiment may turn out to be the most interesting experiment of them all, but so far it has been hard to undertake due to the lack of a strong CW UV laser source and fast beam valves for a rapidly pulsing source.

Photoelectron spectroscopy can be considered to be a back titration; you illuminate the sample with photons at 21 eV and back titrate the electrons. This has two features.

1. It requires an accurate and calibrated electron monochromator, which involves the measurement of absolute energies, usually by employing a calibrating gas.
2. It requires slits or timing for high resolution. The ultimate resolution is usually limited for reasons that remain only partially understood. Typically the limit is near 10 meV (some 80–100 cm^{-1}).

The easiest way to do these PES experiments today is to measure the electron arrival times in a μ-metal-shielded graphite-coated electron flight tube. The accuracy of this method of measuring electron energies is as good as or better than that of any other known electron monochromator. The photo-ionization efficiency (PIE) curve is given in Fig. 10.3(a). As the energy increases the current measured increases sequentially, like a staircase function; at high energies, however, this relationship becomes extremely complex. Nevertheless, it has been very important historically for measuring ionization potentials with high accuracy, as shown by Watanabe[97].

In threshold measurements, one turns the question around. After excitation, whenever one reaches an eigenstate of the system, the electron detaches with no excess kinetic energy because all the energy has been consumed in the optical transition and little or nothing remains.

What is the state of the art of high-resolution photoelectron spectroscopy experiments? A state-of-the-art PES spectrum with a resolution of 2–3 meV is shown in Fig. 10.4, in which the rotations of nitric oxide are indeed resolved. For this purpose, one goes to a high rotational state of the ion for which the rotational states separate clearly as the energy increases. If you go to high enough energies, the separation of states is sufficient to provide good rotational resolution. The case of nitric oxide is a good example for photoelectron spectroscopy since it is possible to separate even the $N^+ = 22$ state. This assignment can sometimes be difficult unless the results of an *ab initio* calculation are available. This is perhaps state-of-the-art PES today. PES will continue to be of considerable importance in spectroscopy since it is a facile survey method of molecular states that allows one to determine the wavelength range required by the laser system in a ZEKE experiment.

Now it must be emphasized that photoelectron spectroscopy was of enormous importance. It had a great impact on our whole knowledge of

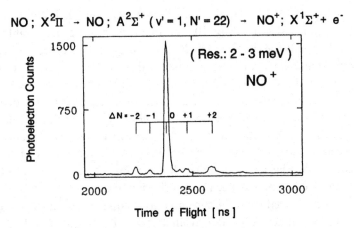

Fig. 10.4 The high-resolution photoelectron spectrum of nitric oxide at high rotational energy[98].

energy levels, electronic energy levels and photoelectron cross sections. Complete books have been published on various cross sections. This tradition has continued from the book by Turner *et al.*[65] to that of Kimura *et al.*[99], which collectively contain many extremely important photoelectron spectra. Much of what is known about energy levels is due to the wealth of papers on photoelectron spectroscopy.

Photoelectron spectroscopy also is involved with electron energy analysis, for which resolution typically decreases with energy. Since PES monochromators maintain a constant resolving power of $\Delta E/E$, as E increases, the resolution decreases. Also the absolute inaccuracy, if one tries hard, can be kept down to some 40 cm^{-1}. It does not appear to matter what type of photoelectron spectroscopy is employed. Today, PES electron monochromators curiously all seem equal in having a limiting resolution of about 10 meV (80 cm^{-1}), although EELS does better. This even appears to be true for time-of-flight electron analysis. The fact that one has certain stray fields (typically 10–20 meV cm^{-1}) in the apparatus also causes problems since an electron travels 10 cm in a field of only 10 mV cm^{-1} in 1 μs. These effects, therefore, are quite severe for very slow electrons and have to be taken into consideration. Surfaces are equally problematical. Different oxide surfaces produce various surface potential domains. Colloidal graphite or gold is often employed although the benefits are hardly measurable. The two difficulties we wanted to address in photoelectron spectroscopy were the need for absolute calibration of the spectrometer and its limiting resolution.

Our first study was based on a new *Ansatz* employing threshold steradiancy detection as shown in Fig. 10.2(a). In a certain sense, one might argue that this is already the situation in which photo-ionization detection is done with zero-kinetic-energy electrons, albeit without delay. The ion detected in coincidence with this threshold electron is at the energy state defined by the electron threshold. Of course, if one has the state selection on the one side from this exciting electron, then one has the great virtue of state selection for the positive ion on the other side. One can carry out experiments with the ions in this selected state, either for unimolecular reactions or for state-selected bimolecular and ion–molecule reactions. This is a very good way to characterize by coincidence the positive ion and its energetics uniquely. This method, derived from the threshold technique, has been used to great advantage by T. Baer in North Carolina and by P. M. Guyon in Paris and is known by various acronyms such as PIPECO. It has only just begun to be used with ZEKE state selection[100–102].

For the ion coming out at zero energy, Fig. 10.3(b) shows the simple plot that is involved. The limitation is derived from the light source used. Therefore, the resolution is of the order of two tenths of a wavenumber with present pulsed dye lasers. This resolution could be much better provided that one would use a better light source. In previous work, we have done sub-Doppler experiments with CW lasers and pulse-amplified CW lasers. These experiments can, of course, be extended to ZEKE spectroscopy. The question of what can yet be done for resolution is very promising, which is unfolding at present[37].

One might also ask the following question: what is the principle of threshold detection? This detector, as I pointed out briefly, does not depend on measuring the electron kinetic energy at zero excess energy, i.e. energy is not measured, but rather a different property of the zero-kinetic-energy electron. This difference property reflects a unique steradiancy behaviour and, thereby, a peculiar way of selecting the solid angle subtended by the zero-kinetic-energy electron. As mentioned before, for systems with non-zero-kinetic-energy electrons one cannot fit the electrons into a hole at the end of a long drift tube. This is only possible for zero-kinetic-energy electrons that have no native velocity component perpendicular to the beam. This is peculiar to zero-kinetic-energy electrons. For all other electrons this perpendicular component cannot be induced to undergo a well-behaved transit into the hole at the end of the long channel. That, of course, is the most optimistic way of looking at it. Clearly, there could also be a velocity component along the beam axis against which one cannot discriminate simply. If you look at the transmission functions (Fig. 10.5), this leads to

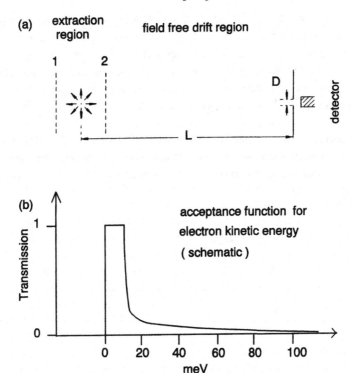

Fig. 10.5 The steradiancy analyser. (a) Only threshold electrons have no perpendicular velocity component and hence make it through the 'hole' at the end of the channel. (b) The transmission function of a simple analyser.

the high-velocity tail. The reason is that the method does not discriminate against the detection of fast, straight-through electrons. Guyon's group in Paris[33,103,104] found one way to solve this problem. By using time-gated detection, one can clip off the non-ZEKE contribution and be able to solve the problem by discriminating against those hot electrons that happen to have a forwards component. The alternative is to employ an electron monochromator at the end of the apparatus as was done by Peatman[34] (Fig. 5.1).

Nevertheless, threshold detection was not optimal in the sense that, although the energy was very accurate, the resolution typically improved by only a factor of two. Although this technique has been of wide applicability to coincidence experiments such as PIPECO it left open the question of how to obtain the resolution promised by modern lasers. What is the reason for this? The problem is that, for any kind of accuracy with respect to the timing of the electron, going down the tube found in Fig. 6.1, which might discriminate against unwanted fast electrons, practically means that you must have picosecond excitation and picosecond timing or an accurate monochromator. Such fast timing in excitation would, of course, seriously degrade the energy resolution. The better solution is to delay the draw-out pulse and thus transform the time resolution required for good resolution into the nanosecond regime, under which a slow excitation pulse can be used together with slow timing that will not degrade the homogeneous pulse width.

The alternative would be, of course, to use femtosecond excitation to increase intensities or to look at very short time scales. This will lead to entirely new and exciting results for large molecular systems, which will be discussed later.

11

Delayed ionization

11.1 The effect of time delay on measurement

The next level of sophistication involves 'waiting'. One irradiates the sample with a pulsed light source (Fig. 11.1), turns off the light source and waits some 2–5 µs before turning on a draw-out field. Waiting encompasses a number of features. When first starting this work, we expected that, if one were to photo-ionize and then wait for a few microseconds, one would lose the signal and see nothing. We did, however, see a signal. There are good reasons for this, which we now know in hindsight. The simplest feature (Fig. 11.1) is that, when one waits, the cloud of photoexcited species blooms out as a function of time and grows in size as it moves down the flight tube. Whereas the prompt timing between the true origin and the near-ZEKE electron is just a little bit different (picoseconds), as a result of the delay it now becomes many nanoseconds. Allowing time for the cloud to bloom before it is detected means that, for the in-line system (the out-of-line system is lost by virtue of the steradiancy), the time scale is grossly amplified. By delaying, one has made a time-to-amplitude converter and increased the ΔT for near-ZEKE states. Hence, the resolution of these ZEKE electrons has been increased. The species at and slightly below or slightly above the transition move apart and one can observe the central ZEKE electron quite readily. This is done in a µ-metal-shielded set-up to keep stray magnetic fields out . One can, of course, turn the question around. If one has these near-ZEKE electrons, then, rather than rejecting them, what else could one do? To provide an answer, we designed channel plates so that they contained a series of anodes positioned as concentric circles. These were used to pick up the near ZEKE electrons, i.e. the electrons whose paths subtend a very small angle exiting from the source. We simply defined the hole at the end of the tube by the concentric rings of the detector and, in this way, we obtained the near-ZEKE information. The

Delayed ionization

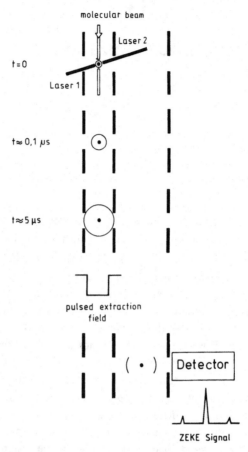

Fig. 11.1 ZEKE measurement by the delayed pulsed field technique. Note the increase in size of the 'electron cloud' of near-ZEKE electrons as it proceeds down the flight tube, increasing the threshold-ZEKE selectivity as it drifts.

near-ZEKE information is, of course, important. It allows one to obtain information about the polarization of the transition and this, of course, is of interest (Fig. 11.2). The experimental results are displayed as diagrams for the various rotational states in the originating S_1 state and this is seen in the ion. The bottom of Fig. 11.2 shows the experimentally observed envelopes. These distributions are cast in the form

$$I(\theta) = \frac{1}{4\pi} [1 + \beta P_2(\cos \theta)]$$

involving the asymmetry parameter[105] β. These simulations (shown in the upper part of Fig. 11.2) were performed by Rudolph *et al.*[72]. The first key

Simulation

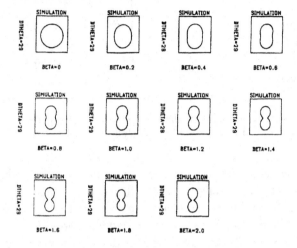

Experiment NO

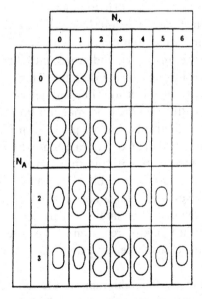

Fig. 11.2 The angular distribution of near-ZEKE electrons using the annular ring
analyser[72].

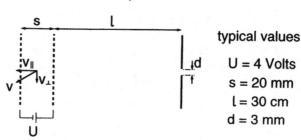

Fig. 11.3 The energy and timing of the static draw-out field versus delayed pulse mode. Typical values are $U=4$ V, $s=20$ mm, $\ell=30$ cm and $d=3$ mm.

idea is then the delay between excitation and extraction. The delay brings about two things: by amplifying a ΔT it spreads out the arrival time differences for the on-line electrons and also it allows there to be a drift time for the slightly perpendicular electrons. The latter phenomenon can be used to measure these near-ZEKE states or, if the drift is extreme, the wait eliminates these as hot electrons. This is the simplest effect from the ionization delay.

As a review, the simplest way to demonstrate the time delay method is to have an excitation in Fig. 11.1 at $t=0$ in the source. The laser excitation can involve two wavelengths; the first colour being used for the excitation to S_1 and the second colour for the excitation to ionization. The electron cloud moves down the detector and grows and grows until, finally, only the core of this cloud makes it through to the detector. This core is now greatly separated in time due to the time delay. Thus, one detects within a tight time window. It is easiest to illustrate the large magnitude of this effect by considering a short 20 mm flight tube and a 1 μs delay in Fig. 11.3. Consider the detector in an in-line and in an out-of-line configuration. $v_\perp$ shows that the maximal wrong energy that is transmitted with a 1 μs delay has decreased to 0.03 cm^{-1} as shown in Table 11.1. Table 11.1 also shows that an offset of 0.1 meV gives a time spread of only 0.04%, against which it is hard to discriminate. On the other hand, the 1 μs delay amplifies the time difference to 17.8%, hence making rejection of the false energy with time gating quite easy. In the case of the static analysis, one has very poor discrimination.

For ZEKE spectroscopy the achievable resolution increases by 2.5–3 orders of magnitude. To illustrate this point look at the original spectrum from the book by Turner *et al.* on nitric oxide (Fig. 8.5), which shows the $v^+=0$–4 vibrational transition in the inset above. We expand the $v^+=0$ state from the spectrum above and take a 3 meV slice that is represented by about

Table 11.1. *Effect of time delay on resolution. (a) $v_\perp$ initial perpendicular velocity component ($v_\perp(max) = d/(2t)$) and (b) $v_\parallel$ initial parallel velocity component (see Fig. 11.3)*

(a)	Flight time t (ns)	Max detected vertical component $v_\perp$(max) (m/s)	$E_\perp$(max) (cm^{-1})	
Static	381.5	3.93×10^3	0.35	
Delayed (1 μs)	1381.5	1.09×10^3	0.03	
(b)	Flight time t (ZEKE) (ns)	Time separation from near-ZEKE peak^a (ns)	t (0.1 meV) (ns)	Separation as percentage of total flight time (%)
Static	381.5	381.7	0.2	0.04
Delayed (1 μs)	1381.5	1313.7	67.9	17.8

Note:
^{a} A near-ZEKE peak is here defined as one that is $\Delta E = 0.1$ meV away

the width of the line drawn. When this is expanded to the spectrum in the circle below, the rotational structure is revealed at the origin and also for higher levels of rotational N^+ states. When one repeats this type of spectrum for the benzene cation (Fig. 11.4), matters look a little worse; first, the spectral resolution is 1 cm^{-1}. The rotational level structure is nearly observable, which is the best which our ZEKE apparatus could do at that time (*vide infra*). We could just see the onset of rotational structure in the early days, but this has been improved over the years until now detailed rotational structure can be resolved (Fig. 11.5).

11.2 The observation of long-lived ZEKE states

The systematic problem we had at the start of our work has turned out to be quite a blessing. Table 11.2 summarizes the important facts at that time. Particularly of interest is the ionization potential for nitric oxide, which is the traditional benchmark. This is perhaps the best known IP of all due to the beautiful work of Miescher *et al.*[30] and Miescher[42]. He obtained a value of 74 721.5 cm^{-1} for the IP and before we discovered the long-lived ZEKE states we obtained 74 717.2 cm^{-1}, i.e. we were some 4 cm^{-1} below the correct value. Now 4 cm^{-1} is not much in a historical context, but it was enough to

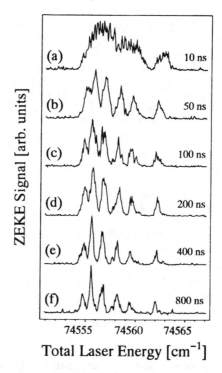

Fig. 11.4 The onset of rotational resolution in the $^2E_{1g}$ vibrationless ground state of the benzene cation. The ZEKE spectrum for the $(C_6H_6{}^+)\,^2E_{1g} \leftarrow (C_6H_6)\,^1B_{2u}$ transition obtained in a two-colour experiment in a skimmed jet system with the $J' = 2$ rotational level in the intermediate state populated. The resolution is increased when the ionization pulse is applied as a slowly rising ramp. The peaks are assigned to rotational levels of the benzene ion as (from the right-hand site) $N^+ = 6$, $N^+ = 5$, $N^+ = 4$, $N^+ = 3$, and $N^+ = 2$, 1 and 0 (unresolved). Note that the full width of the structure is about 1 meV[44,106].

cause some concern. We had the same problem for benzene, for which our result was too low by 17 cm^{-1}, but in that case the Rydberg extrapolation turned out to be inaccurate. Hence, we now feel that the ZEKE IP is better than the extrapolation from poorly resolved Rydberg states. This has to do with the difficulty in observing a complete Rydberg series, particularly for the rotationless origin of each state under moderate resolution. So one needs to be careful when deciding which method is better. All our extrapolations for NO were systematically too low by a tantalizingly small number, which could not be understood but which turned out to be the key to the present understanding of ZEKE spectroscopy.

Of course, stray fields can also be quite a problem, since zero-kinetic-

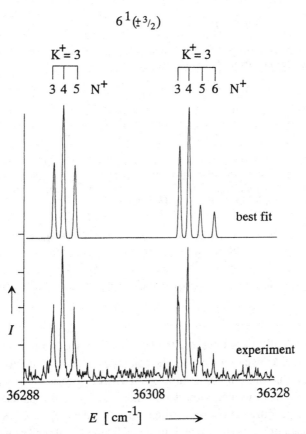

Fig. 11.5 Jahn–Teller splitting in the benzene ion. The lower trace is the experimental spectrum; the upper trace is the best fit to the calculated spectrum[107].

energy electrons are very sensitive to small electric stray fields, as seen in Table 11.2. In particular we had a stray field of some $20 \, \text{mV cm}^{-1}$. Also, an electron in a $20 \, \text{mV cm}^{-1}$ field travels some 18 cm. Hence, one has to worry about these difficulties and therefore one needs to control stray fields. Contrary to what we expected from such low fields, we noticed very little change in signal strength no matter what the extraction delay. This also caused us some concern. Biernacki *et al.*[43] from Yale re-measured the excited A state of nitric oxide and found nothing wrong with the original assignment. We had hoped that the old assignments were wrong, but they were correct. Also Miescher's work was highly reliable. This discrepancy in the IP values was real, but we could not explain it. We found the key to the problem in a very sophisticated experiment by one of our undergraduates,

Table 11.2. *Intriguing facts – old data*[108]

- IPs determined from ZEKE experiments are systematically lower than values determined from Rydberg extrapolations

	ZEKE-PES (cm^{-1})	Rydberg extrapolation (cm^{-1})	Reference	ΔIP (cm^{-1})
Benzene	74555.8±0.5	74573.0±2	45	17
Nitric oxide	74717.2	74721.5	42	4.3

- Sensitivity to stray electric fields: ZEKE electrons are extremely sensitive to small electric fields during the 'field-free' delay

for a delay of 1 μs:

stray field:	20 mV cm^{-1}	⇒	electron travels	18 cm
	10 mV cm^{-1}	⇒	electron travels	9 cm
	2 mV cm^{-1}	⇒	electron travels	1.8 cm

- There is essentially no loss of signal on going from static to delayed (1–5 μs) extraction

Reiser, who really found the key to the issue in a set of three very simple experiments taking cognizance of the small field. This result turned out to be the key to the present understanding of ZEKE spectroscopy, at least for positive ions. After we had understood this effect, all discrepancies became clear and procedures to eliminate them could be developed. ZEKE results became as accurate as or better than those from any other known general technique for measuring ionization potentials (Table 8.1).

Now, what is this experiment? This experiment, as shown in Fig. 11.6, looks at the ionization of nitric oxide under three quite similar conditions: (a) slightly below the ionization potential, (b) near the threshold and then (c) still higher above the threshold. Just below the threshold (a), the time of flight of the electron appears to be independent of the delay time employed. It does not matter when one draws the electrons from the nitric oxide out of the apparatus, since it seems that one need not care about the time. Hence, it is clear that, whatever this peak is, it cannot originate from free electrons because their behaviour should depend on the stray fields as well as on the time delay that one employs. It is not a free-electron effect. On the other hand, one is also operating below the ionization potential. Hence these are not free electrons, but become free electrons in the field. Let us go a little higher in energy to (b), just above the known ionization potential. Quite clearly, one now has free electrons. The peak again appears but now

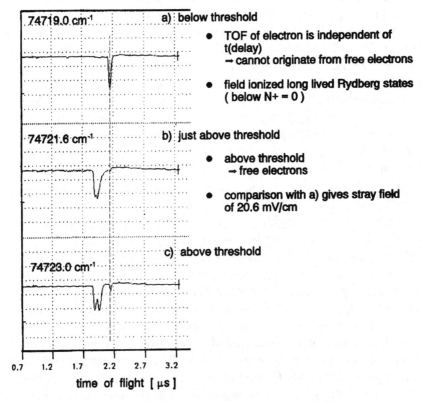

Fig. 11.6 The ionization energy of nitric oxide[35].

it changes without an applied field to a different time of 1.88 μs, i.e. it leaves the extraction region without the electrons being subject to an applied field. Hence, these are now electrons that must have experienced a stray field of approximately 20.6 mV cm⁻¹. Notice that there is still a small peak from neutral species as in (b). If one goes a little higher in energy above the threshold to (c) one really sees the peak as well as the same slight signal from the neutral species. The peak is split due to the forwards and backwards velocity components of the electrons. What this is telling us for the very first time is that we are producing neutral ZEKE states in addition to the free-electron states. These are peculiar Rydberg states that are very long-lived. In fact, we had already speculated in the past about the effect that long-lived Rydberg states may have on ZEKE spectra. The estimates we could make for the lifetimes of these Rydberg states, however, were of the order of 20 ns. If one were to extrapolate with the well-known n^3 law, our lifetimes would be at least two orders of magnitude too long for Rydberg states. An n^5 dependence[109] for which there is a sound reason, would reduce this discrepancy. This has been confirmed by the recent experiments of Even *et al.*[110] in Tel Aviv. Thus one finds an unexpectedly long lifetime. This was the key experiment which for the first time demonstrated that we must take cognizance of the extreme lifetimes of ZEKE states, which are, of course, neutral species within a very narrow band below an IP at very high *n*. One measures here just below the onset of ionization and obtains very long-lived neutral Rydberg species that are then field-ionized, just above the ionization potential. The charged species can migrate in the presence of a field, but this neutral ZEKE state is still in place. This experiment caused the pieces of the puzzle to fall into place. This lifts the mystery of the old low IPs with the wrong threshold method. Knowing this, one can now extrapolate and obtain highly accurate IPs as seen above (Fig. 11.7). The longest lifetimes of these Rydberg states are still those for the states with highest *n*. Considering that this delay can be some microseconds, Reiser performed a field extrapolation and obtained an IP of $74\,721.7$ cm^{-1}, in perfect agreement with the Rydberg extrapolation. The resolution of the apparent discrepancy was the discovery of the anomalous ZEKE lifetime effect, which became the key for the new spectroscopy.

These Rydberg states are detected with a draw-out pulse, but the nature of this draw-out pulse is also highly critical. This is an important problem with the draw-out pulses. One could impose a sharp draw-out pulse (top left-hand part of Fig. 11.8) and compare this with the slow draw-out pulse originally discovered using a poor pulse generator. It gave a reasonable spectrum. Then we decided to improve this terrible pulser, in order to obtain a good pulser and hence to do a better job (top left-hand part of Fig. 11.8). It should really pull the system out even better because of its very fast rise time. This conclusion was totally incorrect; in fact it led to worse spectra. One apparently must use not a fast pulse generator but rather a slow pulse generator. What is the reason? The reason is exactly the same as before. Here the entire system is seen with the pulse at the same time and all the Rydberg states as well as all the direct charges or electrons are seen simultaneously. As a result, a very broad spectrum is produced for all these species. One clearly must distinguish among the various cases. First a small field is present that ejects all the direct electrons or direct ions and, then later on, a gate is set for the field-ionized species, whereupon electrons from

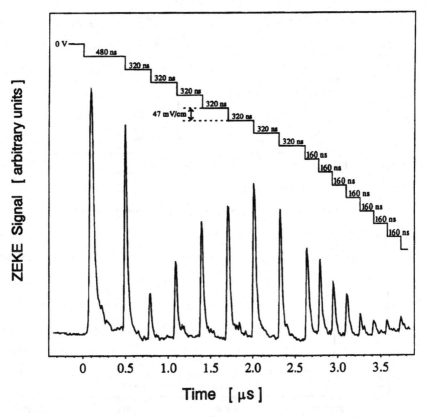

Fig. 11.7 Stepwise pulsed-field ionization for zero-field extrapolation.

about 2.8 cm^{-1} below the IP are collected. Under these conditions, a highly resolved spectrum (the bottom part of Fig. 11.8) is measured. The early portion of the slow pulse ejects the direct ionization and the later part of this pulse ejects the ZEKE state. It is the highest Rydberg state that has formed a ZEKE state which gives this optimal resolution. The resolution will thus become atrocious for fast pulses due to a pulling in of lower Rydberg states. This was the case with the fast pulser. The detection is optimized for the highest neutral Rydberg state just below the IP or other relevant ionic state. These are states that 'hug' the ionic state from below and belong to the species detected in cation ZEKE spectroscopy. Present high-resolution experiments thus require a programable pulse generator or pulse slicing. With this technique it is possible to obtain high-resolution ZEKE spectra. The primordial spectrum is shown in the bottom part of Fig. 8.5 together with the PES spectrum of the first vibration.

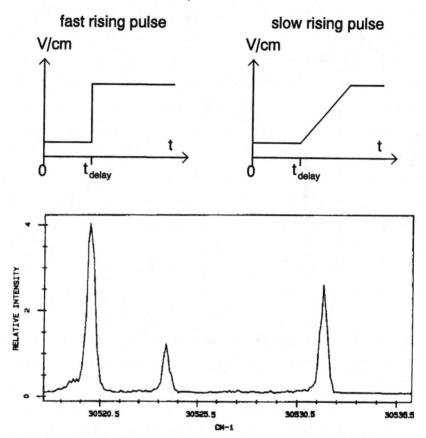

Fig. 11.8 Slow and fast delayed-field extraction. The fast rise gives a poorly resolved spectrum. The slow pulse allows the ZEKEs of higher energy to arrive before those of lower energy, thus increasing resolution as shown in the spectrum below[35].

In summary, waiting in a zero-field environment is a very important feature that not only produces highly accurate results but also generates a whole series of interesting effects, all of which are fascinating, so that they should be kept somewhat apart. Waiting is useful, but it encompasses a number of phenomena.

1. It excites systems in near zero field, which is of classical importance in spectroscopy.
2. It amplifies the arrival time differences for the in-line electrons and, hence, greatly increases the spectral resolution.

3. It allows the perpendicular electrons to disappear by increasing the extent of the cloud of charges with time.
4. The wait allows all charges to disappear. Then only the ZEKE neutral species are observed, through applying a fiduciary field and detecting the 'last' Rydberg state, which is hugging the eigenstate from below. Thus, one can make measurements arbitrarily close to the IP.

11.3 Extrapolation with applied fields

Finally, one can refine the experimental system significantly. This can be easily seen by showing a staircase function (Fig. 11.9) for the draw-out voltages. One can divide the voltage ramp for the draw-out pulse by programming a pulse generator to act as a staircase. This puts in more and more electric field, sequentially adding to the field, so that for each step one can measure the associated current measuring how electrons are produced by each step. The onsets of a series of peaks derived from every small step of $47 \, \text{mV cm}^{-1}$ can be observed. One can extrapolate back to zero field and obtain the diabatic value for the field dependence of $3.86\sqrt{F}$, which is in fact observed. One can see the peaks by looking at the threshold at the IP. When the voltage decreases slightly, the next peak is observed. Thus, one can actually scan the peaks for each step of the entire ZEKE energy spectrum near the IP. Notice here that peak intensities for the various steps do not just change monotonically, as one expects for a spectrum, but rather show important structure. It should be noted again that this value subsumes excitation from high-n mixed states above the Inglis–Teller limit where this mixing is produced by the stray fields present in the laboratory. This, however, represents the typical conditions for ZEKE spectroscopy. The word diabatic is confusing here. The transition has really a slow slewing rate more typical for an adiabatic transition, but for high-ℓ systems, with a small quantum defect, the system will behave diabatically due to the tight spacing of the levels and thus exhibit a $3.86\sqrt{F}$ behaviour. All practical ZEKE measurements appear to be within this domain.

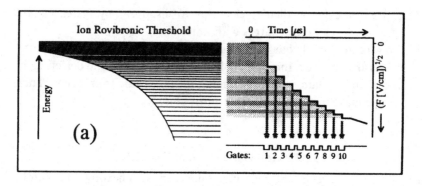

BENZENE: $D_0 \, {}^{ev}E_{1g}(N^+,K^+) \leftarrow S_1 \, 6^{1 \, ev}E_{1u} \, (J'=4,K'=4,+l)$

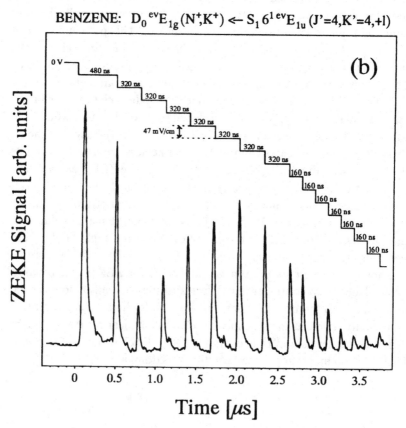

Fig. 11.9 PFI-ZEKE with a 'staircase slope' pulse[108].

12

The mechanism

It should be mentioned at the outset that the discussion here draws not just on the work in Munich but also heavily on the pioneering work of Gallagher[111] and Mahon et al.[112] for atoms and on that of Bordas et al.[113,114] and Chupka[62]. The theory has been brought forth by many groups, principally the groups of Levine[115–117] and Jortner[118], but also by Merkt et al.[119–121].

ZEKE states are related to Rydberg states from which they originate from the low-ℓ optically populated states, to states high into ℓ and m_ℓ to finally producing a spectator electron that is weakly coupled to the core and hence has an enormous lifetime. This facile migration in ℓ space to a long-lived spectator state was quite a surprise. Clearly, the ZEKE state originates from a specific Rydberg state. It is equally clear, however, that it has an existence of its own. In fact, it was shown recently by Alt et al.[122] that these ZEKE states can, in principle, be produced in the absence of optically excited Rydberg states. Hence, in principle, their connection to Rydberg spectroscopy can be quite tenuous.

The current explanation for the longevity of these states is seen in high-n Rydberg states originally populated by photons in low-ℓ states having strongly parabolic orbits. In this state the proximity to the core goes as ℓ $(\ell+1)$ and thus, for low ℓ, it is particularly sensitive to external fields. In the presence of a small stray Stark field, such as will be present in all laboratory experiments, these parabolic Stark states exist and already contain a set of ℓ states. As ℓ increases these orbits become more circular and the electron tends to stay further away from the core. This separation from the core goes as ℓ $(\ell+1)$. Correspondingly, the quantum defect becomes small as ℓ increases and hence the orbits become hydrogenic. However, only near the core can the electron change its state. An electron far from the core cannot even absorb light. Hence the electron in a high-ℓ state becomes a spectator with respect to the ionic core and its lifetime becomes anomalously long. Among the Rydberg states, one also produces some low Rydberg states that decay so rapidly that one does not see them. One then has a band of these higher ZEKE states and, within this 'magic' region with the very highest n states, one can impose a small spoiling field to eject any direct ions and extract the middle slice in the magic region to form the ZEKE spectrum. Of course, the same holds for those ZEKE states above the ionization potential which exist as islands of stability far into the continuum. Since the ZEKE states are slightly below the eigenstates of interest this typically leads to spectra with a very small decrement in the ionization potential of 1–2 cm^{-1}. This decrement could either be absorbed into the calibration or measured and extrapolated from two or more potentials, as was done when we applied multiple voltages in a staircase shape in the draw-out pulse for extrapolation. In this way, one can actually get rid of this decrement entirely.

The puzzling feature not answered here is still the extreme 20–100 μs lifetime of ZEKE states. This fact has been confirmed independently by Bahatt et al.[115], Krause and Neusser[123], Scherzer et al.[124,125] and Even et al.[63] among others, even for more complex molecules. It is interesting to ask what mechanism is involved in the migration of the ℓ quantum number. Several mechanisms have been proposed to explain the electron's entering a ZEKE state: one is a rotation–electron coupling[2] and another is the theory due to Bordas et al.[126] and Chupka[62] which suggests that the ubiquitous stray fields interact in a Stark manifold to couple these states. A further possibility is that a combination of both the foregoing applies.

The stray field was determined in some experiments[127] to be 40 mV cm^{-1}. The 8 cm^{-1} band of ZEKE states for this is shown in Fig. 12.1. This band remains the same in Fig. 12.1 if the field is increased to 200 mV cm^{-1}. Hence one can refer to this as the low-field region. The signal remains the same throughout this region. This signal is even the same irrespective of whether the field is on with the laser pulse (in-field) or delayed (out of field). The value 40 mV cm^{-1}, however, is above the Inglis–Teller limit, so all states are mixed by the Stark field. Hence the conclusion here is that, once the field has mixed all the states, no further effect will be caused by the field. The field produces mixed-ℓ Rydberg states, which also occurs with very-high-resolution spectroscopy[37,128]. These are not yet ZEKE states, which also demand m_ℓ mixing. Vrakking et al.[129] have shown that, when one adds a lot of ions with a third laser, more ZEKE states are obtained for xenon. Alt et al.[130] found no effect with benzene. The question of the function of ions in molecular ZEKE spectra then arises. Recent work has shown that, if the

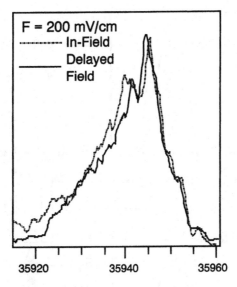

F = 200 mV/cm
In-Field
Delayed
Field

35920   35940   35960

Fig. 12.1 The dependence of the ZEKE signal width on spoiling field strengths. The top panel shows ZEKE spectra recorded using a 200 mV cm^{-1} spoiling field. The field is present during excitation in the spectrum shown as a dashed line and is delayed by 100 ns in the spectrum shown as the full line (field-free excitation).

concentration of benzene ions is drastically reduced below the typical conditions used by Alt *et al.*[130], then one clearly sees a dependence of the ZEKE signal on the ion concentration also for benzene (Fig. 12.2). One can now compare the ion concentrations at a typical voltage in the ZEKE band of some 1 V cm^{-1}, both in field and out of field. The former would influence the mixed-ℓ Rydberg states, whereas with the latter the ZEKE states are not so affected. The ratio of the ZEKE signals produced by these ion concentrations is presented in Fig. 12.3. Here one can clearly see the dependence on the ion concentration when producing ZEKE states from mixed-ℓ Rydberg states. Note also that the graph intersects near zero, indicating again that fields in the absence of ions do not form ZEKE states. Hence ions are required in order to produce ZEKE states.

The function of these ions is to establish an equilibrium between Rydberg and ZEKE states quickly under normal conditions, as shown by Alt *et al.*[130]. The entropy then drives the system to the higher state density of the ZEKE states.

The ZEKE state, therefore, can be viewed as having been produced from a mixed-ℓ Rydberg state, i.e., the core of the state has an electron sitting in a near-circular orbit with at least a reasonably high ℓ quantum number.

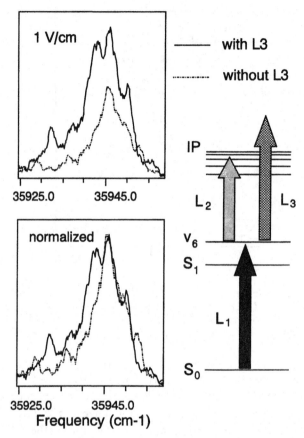

Fig. 12.2 Comparison of ZEKE spectra with (full line) and without (dashed line) additional ions produced with an additional 266 nm laser. The introduction of the 266 nm laser led to a twofold enhancement of the ZEKE signal. The presence of the third laser, producing additional ions, also caused an increase in the spectral width, as can be seen in the bottom panel, in which the intensities of the ZEKE signals with and without the third laser are normalized.

These Rydberg states are now converted by long-range ion interactions into mixed-m_ℓ ZEKE states with increased intensity. Here the electron essentially becomes a spectator system with respect to the core. Indeed, the electron becomes a spectator in an interesting way since this spectator-like property with respect to the core is extreme. These states (Fig. 12.4) exist below the eigenstate of the ion, thereby forming 'hugging' states, which we call ZEKE states. These are produced from optically accessible, high-energy Rydberg states and eventually lead to the ZEKE process that forms the long-lived states. They are, however, distinct from these optical Rydberg

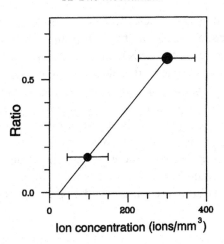

Fig. 12.3 A graph of the integrated PFI signal intensity, for which excitation occurs in the presence of a 1 V cm^{-1} spoiling field, divided by the integrated PFI signal intensity, for which the same spoiling field is delayed by 100 ns as a function of the ion concentration.

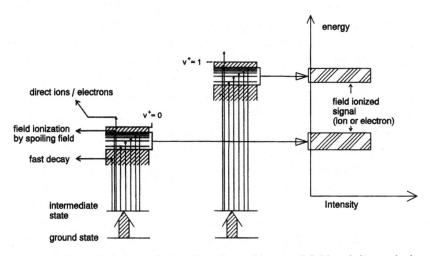

Fig. 12.4 The technique spoils the direct ions with a small field and then waits in order to eliminate low-n states and obtain high-n ZEKE states. Note that some ZEKE states are lost in the spoiling field.

states and, in fact, can be produced indirectly without their optical precursor. The independence of the spectator electron is such that one can even have photodissociation of the core without affecting the electron. Interaction with the core is very strongly uncoupled but, of course, not totally uncoupled. We can even prepare ZEKE states above the classical ionization potential, above which the blue states, if hydrogen-like, would not ionize, neglecting ionization by tunnelling. For molecules with a quantum defect, however, the ZEKE electron has some residual coupling with the core and the lifetime of this electron is a measure of the coupling strength. If one prepares such a ZEKE state and then puts it into the ionization continuum by dropping the energy with a field, one finds that it still takes typically 37 μs for this long-lived blue electron to couple to the core and ionize[131]. This coupling with the core in turn depends on the proximity, which in turn decreases with the m_ℓ state populations. Hence the ions operate to increase lifetimes via m_ℓ, rather than collisionally decreasing lifetimes, again demonstrating the ion-induced migration to higher m_ℓ.

13

Ionization potentials from ZEKE spectra

The various values of ionization potentials that are obtained with various methods for benzene are shown in Table 13.1. There are five entries from our own laboratory with our most recent results having a precision of within ± 0.3 cm^{-1}. The Rydberg extrapolation (bottom line) can be seen to be off by as much as 17 cm^{-1}, although this can be eliminated if one goes to sub-Doppler spectroscopy[4]. The usual textbook wisdom is that a Rydberg extrapolation is the optimal way of obtaining ionization potentials. This is correct for hydrogen, but for few other systems. The reason is quite simple. If you have sub-Doppler resolution, you obtain the rotational structure and, hence, you know what the rotational origin is for each of the Rydberg transitions. Under these conditions, the method works very well, providing that high-n states are not otherwise observed. This procedure has been quite difficult, if not impossible, to utilize for large molecules up to now. Diatomic molecules are good candidates for a Rydberg extrapolation.

Table 13.1. *Comparison of ionization potentials (IPs) for benzene*

Method	IE (cm^{-1})	Precision (cm^{-1})	Reference
ZEKE[a]	74556.1	± 0.3	Lindner et al.[108]
ZEKE[a]	74555.0	± 0.4	Chewter et al.[44]
MATI[a]	74555.5	± 0.5	Krause and Neusser[79]
MATI[a]	74556.0	± 0.5	Németh et al.[132]
PES	74536.0	± 120	Åsbrink et al.[133]
PES	74605.9	± 400	Kimura et al.[99]
2C-MPI[a]	74528.0	± 15	Duncan et al.[47]
2C-MPI	74528.0	± 26	Fung et al.[49]
Rydberg extrapolation	74573.0	± 2.0	Grubb et al.[45]

Note:
[a] Results from the author's laboratory

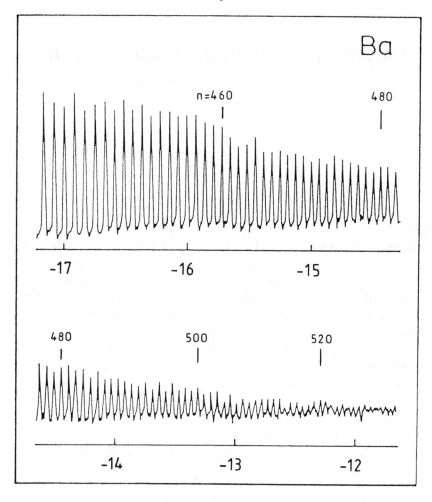

Fig. 13.1 Rydberg states of barium. This demonstrates that high-*n* states can be observed in the laboratory[134].

In general though, not many Rydberg transitions have been identified for complex systems and certainly the rotational structure is rarely seen. With ZEKE spectroscopy, one can obtain very highly accurate values of the IP (Table 13.1). In fact, the reverse has been done, in that Rydberg extrapolations are now often checked with ZEKE results.

The question of analysing high Rydberg states, of course, is not new. Rydberg states can go high in the *n* quantum number. In Fig. 13.1 this can

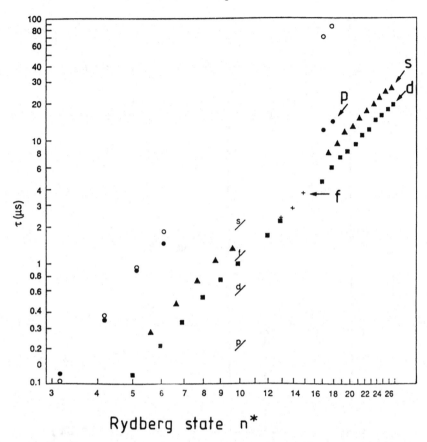

Fig. 13.2 Na radiative lifetimes versus n^*. Experimental values for ns (▲), np (●), nd (■) and nf (+) states are shown. The lifetimes of the states below $n=15$ have been measured by fluorescence techniques, at temperatures of approximately 400 K. The lifetimes of $n>15$ states have been measured by field ionization, the ns and nd states at 30 K and the np states at 300 K. The theoretical 0 K np lifetimes are also shown (○). They are far above the measured values at high n due to black-body radiation. Finally, the hydrogenic lifetimes are shown by the line segments (—)[135].

be seen for barium in the spectrum of the 6s nd ^{1}D$_2$ Rydberg series with a stray electric field below 45 μV cm^{-1} during excitation[134]. The Rydberg states were ionized as a result of collisions with other barium atoms and are visible with $n=520$. In fact, barium clouds in interstellar space show barium atoms in Rydberg states with n as high as 790. These are indeed huge sized states, here some 66 μm in diameter. If you look at the lifetime of the Rydberg states of atomic sodium, they are well-behaved states (Fig. 13.2).

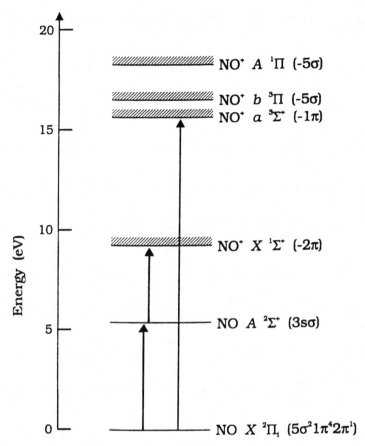

Fig. 13.3 Vacuum ultraviolet ZEKE spectroscopy[136].

Plots of the radiative lifetimes go very nicely as n^3. For molecular ZEKE states, the lifetime increases by two orders of magnitude and then the states become very long-lived. The question one can ask, of course, is that of how long and how highly energetically can one measure with ZEKE spectroscopy. Measuring higher and higher energy ZEKE states into the ionization continuum, one wonders how far these ZEKE states with long lifetimes can be embedded in the ionization continuum before they sense their excess energy and auto-ionize rapidly. They apparently do not ionize, in contrast to the usual textbook situation for the molecular eigenstates, which are deemed to ionize within femtoseconds upon entering the ionization continuum. Such ZEKE work has been done by a number of different groups. In particular, Hepburn's group in Canada has examined at quite high excitation energies (Fig. 13.3) of up to 18.6 eV[136] the third electroni-

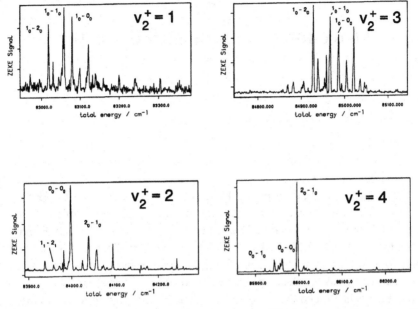

Fig. 13.4 The rotationally resolved ZEKE spectrum[137] of the umbrella modes of ammonia cation.

cally excited state of the nitric oxide ion and observed a very powerful ZEKE spectrum. Certainly, the signal is not quite as intense as it is at the threshold (by perhaps a factor of ten), but a very strong ZEKE spectrum for the electronically excited states remains. In fact, such ZEKE spectra are seen everywhere at all these energies; whenever expected they are seen. This is by no means obvious. It indicates that, even though the energy excess above the IP becomes very large, the Rydberg electron in the ZEKE state does not fly off. Hepburn obtained these energies by generating VUV radiation in a laser, four-wave mixing scheme. We showed for the umbrella modes of NH_3 (Fig. 13.4), for which excitation as high as 0.5 eV above the IP was used, that ZEKE lines are still observed. Hepburn[136] has gone very much higher (to the 16 eV region) with NO^+ (Fig. 13.3) and HBr^+.

As the excitation energy increases, apparently all the ZEKE states expected are still seen. In fact, the ZEKE electron becomes a spectator so that one can photoexcite the ionic core of the ZEKE state and even cause it to dissociate without losing the ZEKE electron, which now is taken over by the fragment.

A list of molecules and their ionization potentials determined by ZEKE spectroscopy can be found on the web site.

14

Detection of ions in ZEKE spectra

As mentioned before, for every ZEKE electron there is always a ZEKE ion produced. Again these ZEKE ions are formed as a small part of the total current in the Watanabe (PIE) spectrum. This means that the ZEKE ions are always buried within a large signal of direct ions. This is also true for electrons, but here the ZEKE states, while they are still neutral, drift with the beam whereas the electrons drift out with the slightest stray fields. Then later a pulse is applied that converts the ZEKE states to electrons and ions suitable for detection. The same is true for ions, which are typically drawn out at right angles. One lets the ZEKE states drift, but now the heavier direct ions will not separate as readily unless a stronger spoiling field is applied. This, however, is a pre-requisite for this method. Any such spoiling field must remove the direct positive ions or for that matter electrons from the ZEKE states prior to application of the field to convert the ZEKE states into ZEKE ions (MATI). The larger the mass the larger the spoiling field required to strip off the direct ions. The larger the spoiling field the more it concomitantly destroys the ZEKE states.

This required stripping field, in other words, will also destroy the ZEKE states, causing them to ionize. Since these ZEKE states are limited to the highest n states ($n = 150$–300) hugging the ion state from below, they will be successively depleted and destroyed by any applied field such as that required as a spoiling field to strip off the direct ions, thus reducing the intensity for the spectrum of interest. Hence, for large enough masses, one will approach the problem that all the high ZEKE states are destroyed by the field required to deflect the direct ions. This then is a natural limit to this technique. The coincidence method mentioned above does not suffer from such limitations.

Zhu and Johnson[40] demonstrated for the case of pyrazine (Fig. 14.1) that analysing either electrons or ions produces essentially the same spectrum, although the latter typically with less intensity and more noise for reasons

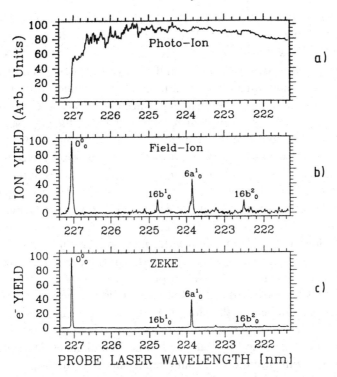

Fig. 14.1 Pyrazine. A comparison between ZEKE with electron analysis (c) and with mass analysis (b). The total photoelectron current (PIE) is shown in (a) for comparison.

explained above. The good news with this method is that one often needs to know with what mass one has been dealing, which could be interesting information particularly if one picks a component from a mixture. On going to mass analysis in Fig. 14.1, one sees stable pyrazine. Pyrazine is not an ideal example for this technique because it does not yet require the mass-selection feature. One does not need the mass information since only pyrazine can be present. Interest in this technique arises for mixtures, particularly those involving unstable species. The total photo-ion spectrum is presented in Fig. 14.1(a), which shows the total current spectrum. The threshold is observable but the signal above the IP is totally uninterpretable. If one looks at the ZEKE spectrum in Fig. 14.1(c), one observes the electron detection scheme in which the various modes appear nicely, just as expected. With mass detection (Fig. 14.1(b)), it is evident that mass detection is completely equivalent to, albeit more noisy than, the ZEKE spectrum (Fig. 14.1(c)). One would expect the equivalence since one can equivalently look at the positive particle rather than the electron of this charge pair produced in the ionization of the ZEKE state. One sees virtually the same spectrum, although with reduced intensity using mass detection. Since the experiments were carried out in two different apparatuses it is difficult to compare these results directly since the mass analysis is much more difficult. This comparison does demonstrate that, since the ZEKE experiment produces an ion together with each ZEKE electron, either the positive ion or the electron, that compose the charged pair, can be examined. Measuring the electron or the positive ion monitors the same process. Detecting the electron signal is preferred in order to obtain good high-resolution spectroscopy with optimum spectral resolution. Detecting the companion cation is of interest when mass selection becomes necessary, e.g. in mixtures.

Typically, in an experiment for ions one has a field for the drift region in Fig. 6.7 followed by field ionization with a high-voltage pulse. To obtain an optimal spectrum, a programmed field is used. Hence, resolution requires careful programming of the draw-out pulse. Basically, when looking at a system, one is not looking directly at the positive ion in the region of direct ionization, but rather at the 'hugging' states just below. For anions, however, one does look at this region. For anions, one looks at free ions on the basis of the assumption that one has no bound states. Whether the dipolar states influence the spectrum remains an open question. This will be an interesting problem to look at with high resolution. Basically, the magic region just below the ionization potential is the interesting point for cation ZEKE spectroscopy. ZEKE spectroscopy is here clearly established also for ions and exploits these ZEKE states in a systematic way even above the IP, where the presence of these many states was totally unexpected.

The mass-resolving technique becomes of particular interest if one has several kinds of chemical system present at the same time. For example, when one has mixed species such as in a cluster jet and wants to use mass selection to detect only the minor constituent. It also becomes of interest when one wants to look at transitional species that have very short lifetimes. The question becomes that of when one wants to examine ions with the added benefit of mass analysis. The disadvantage of this technique is that a substantial spoiling field must be used to eliminate the direct ions. Such a field is necessarily also destructive to ZEKE states. A spoiling field engenders at least two practical problems: it degrades spectral resolution and it reduces intensity. Let us consider a mixed sample. First, benzene again is seen in Fig. 14.2. In more recent work, one can see either the onset in the delayed mode above or the staircase in the total-current mode below. One again sees the splitting due to the quadratic Jahn–Teller effect. This can be done with mass analysis, but it offers no advantage. In fact, it has only neg-

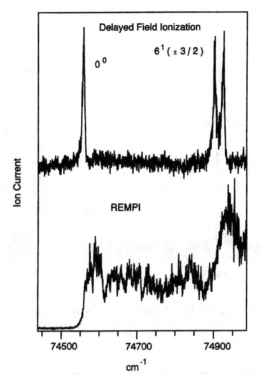

Fig. 14.2 Ion-yield spectra of benzene from two-colour resonant excitation via the $S_1 6^1$ state. Note that the ZEKE spectrum is a component of the total ion REMPI spectrum.

ative effects due to the high spoiling field required, which destroys ZEKE states. We do not need the mass information, since benzene comes directly from a bottle. Something that does not come from a bottle is the benzene–argon complex. A complex of benzene linked to argon is formed in a supersonic jet expansion, which we studied before under high-resolution absorption[138]. This ZEKE experiment is now identical to the pure benzene experiment shown in Fig. 14.2, except that one has switched to a different mass (the upper trace of Fig. 14.3). In fact the benzene spectrum is still there. By just switching from mass 78 to the mass 118 of the complex, one finds a very simple spectrum[139]. The benzene–argon complex formed in the presence of argon within the jet expansion shows a spectrum with the same benzene states as in the identical experimental set-up used for benzene. This is an example in which mass separation provides important new information. Mass selection becomes essential in this experiment.

In Fig. 14.4(a), the standard photoelectron spectrum for benzene is shown. In Fig. 14.4(b), the total current is presented and again a staircase

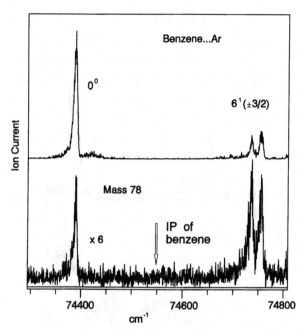

Fig. 14.3 The delayed pulsed-field excitation spectrum of the benzene–argon complex. This demonstrates the ability of mass analysis in ZEKE to work on mixtures[139].

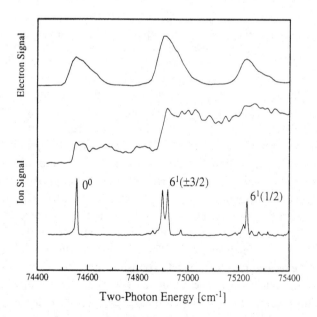

Fig. 14.4 Mass-selective pulsed-field threshold ionization[27]. (a) Shows a normal PE spectrum in a TOF. (b) Shows the total current (PIE). (c) Is the same as (b) but with a delayed draw-out field.

function with some channel couplings of various kinds superimposed is observed. The interesting case is for Fig. 14.4(c), in which exactly the same ZEKE spectrum, albeit at lower resolution, is observed. The interesting aspect is that all three spectra were measured in the same apparatus, specifically, a mass spectrometer in which nothing occurs for a few microseconds after laser excitation before the ions are extracted. One waits a few microseconds before switching the mass spectrometer on. This waiting alone converts (b) to (c) and as a result the structure appears very nicely. Here one sees the splitting of the 6^1 mode into the $\frac{3}{2}$ and $\frac{1}{2}$ components as a result of the linear Jahn–Teller effect of the ν_6 active mode on the $^2E_{1g}$ state of the ion. This is one of four active modes ν_6, ν_4, ν_8 and ν_9. More interestingly, one also sees the splitting of the lower Jahn–Teller component of the 6^1 mode into the $+\frac{3}{2}$ and $-\frac{1}{2}$ components as a result of the quadratic Jahn–Teller effect[79] and in fact even the non-Jahn–Teller-active mode ν_{16} splits into three components (Fig. 14.5) as a result of the Renner effect.

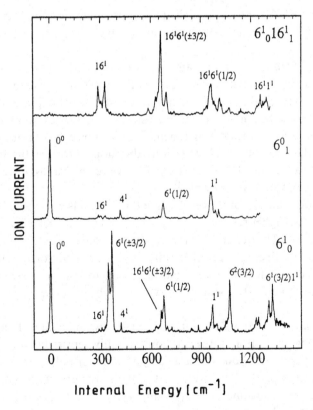

Fig. 14.5 Benzene. Pulsed-field mass-analysed ZEKE spectra of the $\tilde{X}^2E_{1g}$ ionic ground state[140].

15

High resolution – benzene

High resolution in conventional ZEKE experiments with $1-2\,V\,cm^{-1}$ is not possible since one cleans out most of the ZEKE states with a given field. The field must be very small. Even then the width is some $6\,cm^{-1}$ so that, for better resolution, a differential field must be applied to look at just a slice of the ZEKE states. Such slicing is a pre-condition for good resolution. A slight further improvement can be attained by polarity switching, q.v.

ZEKE electron experiments are mostly carried out with dye lasers with a resolution of some $0.2-1\,cm^{-1}$. Clearly the question of whether this is a natural limit to ZEKE resolution can be raised, since it might be expected that there would be inherent limitations due to the stray fields and the attendant Stark effect. Recent experiments[37] have, however, confirmed that, although there are such field limitations, they appear to be in the 100 MHz resolution range ($3\times10^{-3}\,cm^{-1}$). There is, however, also a loss of resolution when too many ions are present[141].

One good example for moderately good resolutions is the benzene spectrum with the higher resolution, which is typical of a ZEKE electron spectrum (Fig. 9.3). The quadratic Jahn–Teller spectrum is observed under high resolution. The lines split completely into two states with rotational resolution. They can, in fact, be rotationally analysed and we can assign the left-hand peaks to the B_{1g} symmetry and the right-hand peaks to the B_{2g} symmetry.

During a visit to Munich Grant assigned this dynamic Jahn–Teller effect. The splitting arises from three rotationally equivalent structures of alternating long and short bonds separated by a barrier of only $8\,cm^{-1}$. The left-hand spectrum of the lower trace is the B_{1g} symmetry state and, for the right-hand spectrum, one has a B_{2g} symmetry of the 6^1 ($\pm\frac{3}{2}$) levels. The dynamical Jahn–Teller effect simply means that one is dealing with a dynamic

118

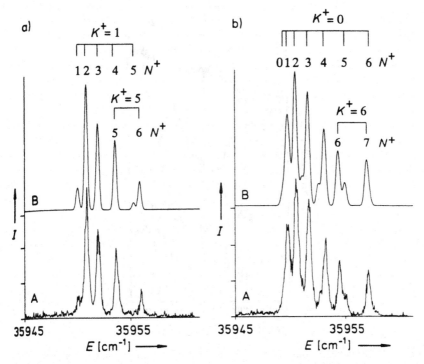

Fig. 15.1 The rotationally resolved spectrum of the benzene cation[142].

motion between these alternating elongated and shortened bonds. The system can be described in terms of a fast dynamic motion between three equivalent structures over a very low barrier and the result shows that D_6 symmetry is still maintained. The resulting splitting is observed and assigned, an effect that is indeed very small (Fig. 15.1). This then represents a good case of a new spectrum of the ion state of benzene that had never been seen before and hence constitutes a first detailed view of the dynamics in the benzene cation.

15.1 Electron versus ion detection schemes

One can drive this equivalence between ions and electrons harder. We have shown that the measurement of ions and electrons (Fig. 15.2) yields similar spectra, as expected. This, however, is at moderate or no mass resolution. Let us return to the ZEKE spectrum of the quadratic Jahn–Teller effect. The total current (a) was measured to show that there is no information

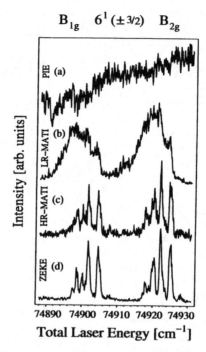

Fig. 15.2 Benzene. A comparison of electron analysis (d) and mass analysis (c) in ZEKE spectroscopy. Note that the spectra are nearly identical, as they must be. The mass spectra are inherently somewhat more noisy. In (a) the total current is shown and (b) uses high-voltage pulse extraction[143].

(PIE). Then one can repeat the measurement with a steep, high-voltage pulse (b) and again such a steep, fast pulse provides little information since the spectrum becomes very diffuse. A slow well-programmed pulse begins to generate a useful spectrum. If one delays by a drift time long enough to perform a ZEKE-like experiment, one gets the ion ZEKE spectrum (c). The electron ZEKE spectrum is seen at the bottom (d). The ZEKE spectrum and the ion spectrum are nearly the same, but the ions produce, of course, a much weaker and noisier spectrum. This then is a trade-off. That you can measure the mass is the good news, but the bad news is, of course, the lower resolution and lower intensity observed.

In the classic ZEKE experiment the ZEKE states produced by the laser are ionized by pulsed field ionization some 200–500 ns after excitation. The direct electrons are dispersed by any small field and by the ensuing flight path to the detector which rejects electrons that are out of line and have the wrong arrival time. The flight of these electrons must be shielded well against external magnetic or electrical fields, or else they will not arrive at the detector. In this method the ZEKE electrons are produced after a short delay and then selected in a well-shielded flight tube carefully trimmed to have extremely low magnetic fields.

Recent measurements have inverted this procedure. Making use of the now known fact that ZEKE states can, under proper conditions, survive for as long as 100 μs, one can produce a beam of ions, electrons and ZEKE states. On letting these drift for some 20 μs, one finds that some 50 mV cm^{-1} is sufficient for eliminating all direct ions for benzene. This voltage does not yet seriously damage the ZEKE bandwidth. At the end of this drift one ionizes to produce either ions or electrons, as desired. Note that, in contrast to historical techniques, here no extraordinary magnetic or electric shielding is required. In fact, a stray field of 50 mV cm^{-1} is desirable and, if not present naturally, must be added. This 'long-beam' unshielded configuration (Fig. 6.7) leads to considerable simplifications in ZEKE measurements and extends the mass-analysed measurements into the kilodalton range[144].

Some caution in comparing electron analysis and mass analysis is called for. At low energies they are naturally equivalent. At extremely high energies, however, we found that the ionic core dissociates while the ZEKE electron remains in orbit. This means that electron detection uniquely measures the original ionic state, even though the core has long since dissociated. These techniques now measure different properties of the system. At very high energies this separation of the ZEKE electron into a spectator orbit means that the electron will still give a proper signal of the original eigenstate of the system, while the ions have long since broken up into many small fragments.

16

Magnetic field effects

In ZEKE spectroscopy, it is important to separate the directly produced ions or electrons from the true ZEKE signal. To achieve this, spoiling via a number of different methods is required. One method involves using a magnetic field. A magnetic field also acts as a spoiler just as the electric field does, i.e., you exploit the motional electric field (Table 16.1) that a moving charge experiences from the magnetic field. The experiment is shown in Fig. 16.1[132]. You put a B (spoiling) field between P_2 and P_3 to extract the ions. The result looks very similar to that with an electric field since spoiling with an electric and a magnetic field have nearly the same effect, as indeed they should (Fig. 16.2). We again obtain near-rotational resolution for the case of the lower Jahn–Teller-split component of the 6^1 ($\frac{3}{2}$) transition in benzene (Fig. 16.3). In fact, you can plot the peak position of the lines as a dependence of the field to show that it is a motional field effect (Fig. 16.4). It is also interesting to notice that lowering of the IP is equivalent to the case of a static field with the same magnitude as the motional field. These combined B and E fields also have an important effect on the lifetimes of the ZEKE states, which also serves to improve the spectrum[145]. The effect of the crossed electric and magnetic fields on the lifetimes of ZEKE states in DABCO (diazabicyclooctane) was the subject of a joint experimental/computational study[145].

Table 16.1. *The motional electric field from a magnetic field*

$F = \gamma\,(vB)$	
F	motional electric field
B	magnetic field
v	velocity of particle
γ	$= [\,1 - (\,v/c\,)^2\,]^{1/2} \approx 1$

Conditions in the jet

$V = 1500$ m s^{-1}	
$B = 0.5$ G	$F = 0.00075$ V cm^{-1}
$B = 100$ G	$F = 0.15$ V cm^{-1}

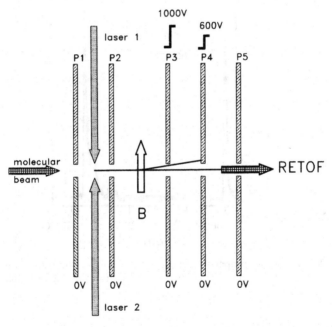

Fig. 16.1 The experimental set-up for magnetic spoiling in mass-analysed ZEKE experiments[132].

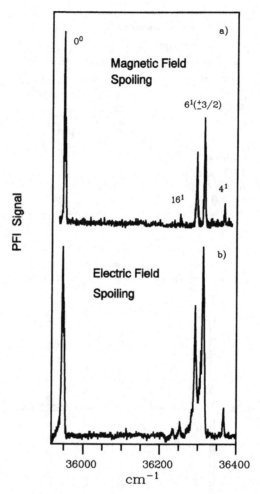

Fig. 16.2 Benzene. A comparison of magnetic field spoiling (a) and the conventional electric field spoiling (b) in mass-analysed threshold ZEKE analysis[132].

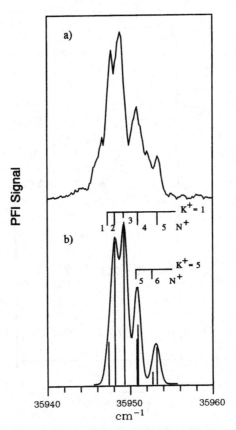

Fig. 16.3 The ZEKE spectrum of the 0^{0+} transition of benzene with magnetic spoiling and mass detection[132].

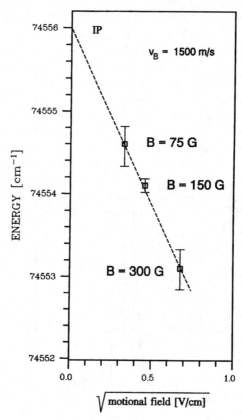

Fig. 16.4 The motional field dependence of the ionization threshold of benzene with magnetic spoiling[132].

17

Anion and neutral species mass-selected spectra

It has been mentioned that one can obtain a routine laser resolution for cations of about 0.2 cm^{-1} and that this resolution is almost 1000 times better than that measured with previously known techniques. The ZEKE resolution is now only limited by the laser utilized. For ions this technique can now be used in conjunction with a series of mass spectrometric techniques. This has some very interesting features. Regarding selection rules, there are significant breakdowns, which make it feasible to observe many, many transitions that otherwise would not be measured were the selection rules rigorously obeyed. The standard example one can give is that of when two species form a complex with a weak bond and one wishes to investigate how many new intermolecular motions are present. One discovers all six new modes. These modes are generated from reducing the number of rotational and translational degrees of freedom from 12 (six from each species) for the two separated molecules to six for the complex. They describe the six new intermolecular motions, although the intramolecular modes themselves are also somewhat modified by complexation. In fact, one can see all six of these modes in ZEKE spectroscopy because new, less rigorous selection rules are operative; that is the good news. The bad news is that up to now it has been very difficult to predict intensities for ZEKE spectroscopy in such a case. Recent results published by Softley appear to provide a theory here[146]. This has its origin in the process of channel coupling. One can see all the intermolecular modes in the complex, for which clearly more general selection rules are in effect and the calculated Franck–Condon factors are extremely low.

One can entertain the observation of negative ions, neutral molecules and radicals. In some very elegant experiments the group of Neumark at Berkeley has demonstrated (starting in 1990) that ZEKE spectroscopy can also be applied to anions[147].

Cation ZEKE spectra have seen the methyl radical[82], ethyl radical[148] and benzyl radical[149] species to date but not the corresponding neutral species. In principle, mass detection for these species can be used at the same time to identify the species definitively while measuring the spectrum. One particularly interesting application is for the ZEKE spectroscopy of Rydberg molecules, species which do not exist in the ground state but are nevertheless of considerable chemical importance. Merkt[150] recently published the ZEKE spectrum of ND_4. This is of particular interest for systems that cannot be put into a bottle. Reactive complexes, clusters, very short-lived intermediates and even transition states[151] can be examined. This method also has some high-resolution aspects that will be developed more fully later.

A brief review of some of the themes discussed for cations may be in order before proceeding into anion ZEKE spectroscopy. Anion spectroscopy has some general features that are quite distinct from cation spectroscopy. The most obvious difference is that there are no Rydberg states that could give rise to ZEKE states. For some anions with a dipole of some 1.6 Debye[152] or higher, however, analogous states exist at much lower energies. These dipole-bound anions were discovered by the group of Brauman[153] and the group of Lineberger[154]. Such anion states typically lie energetically quite close to the detachment threshold and, hence, are difficult to produce with conventional electron attachment, which usually involves hotter electrons. For these anion states, Rydberg electron transfer is often a softer technique[155], after which standard threshold techniques as mentioned before can be employed.

At this stage it is not clear what, if any, resonances exist even closer to the electron-detachment threshold. Such observations have recently been reported for nitric oxide[156].

It remains clear, however, that, even for anions, delayed extraction pulses give a time amplification with respect to the arrival of the electrons, thus increasing resolution. Present resolution limits for anion ZEKE spectra are about 1 cm^{-1}. There are further experimental difficulties due to the large anion clouds that lead to Coulomb repulsion within the cloud and, hence, cause distortions in arrival times.

ZEKE spectroscopy of anions is a new and very promising field and several systems have already been studied with this new technique. The systems studied so far include clusters of metals like gold and silver[157], germanium[158], clusters of silicon[159–161] and carbon[162–164] as well as indium phosphides[165,166]. Also van der Waals clusters of halogens with rare gases[167] and the transition state IHI[168] have been examined. From a chemical point of view this method could also be very interesting for studying free radicals. The spectroscopic information was here obtained from their anion parents, as in the case of OH^- and S^{-169}. The ZEKE technique here particularly allows access to mass-selected unstable molecules. Highly reactive intermediates such as those which occur in metal catalysts have been studied[170,171].

Although the structure of anions is inherently interesting, the most important application of anion ZEKE spectroscopy is to determine the structure of the corresponding neutral species in their ground states. Since anions can be measured using mass selection, the corresponding neutral species are *ipso facto* mass-selected. In other words, this is a method by which a long-sought goal can be reached, namely the goal of ensuring mass selection for neutral species or neutral mass spectrometry. Since the spectra of the mass-selected neutral species are observed, these measurements provide a means of obtaining spectroscopic data for ground-state neutral species, which is particularly useful when they are stable, metastable or in a mixture.

The question arises, why ground-state spectroscopy? Ordinary ground-state spectroscopy is, of course, the proper domain of historical methods such as those embodied in infra-red and Raman spectra whereby the ground state of a particular system is observed. ZEKE spectroscopy also can study the ground states of stable species, but in a rather strange way; essentially by the back door. One can now achieve mass selection of neutral species. One can do this because the anion can be used to produce a specific neutral species. This is of particular interest for neutral species that do not come from a bottle. What does not come from a bottle are species that are unstable complexes, radicals and highly reactive or highly excited species. These systems are observed via unusually intense overtone spectra that result from different selection rules. All are produced from anions.

Anions themselves are of some inherent interest and can be looked at as well. You can measure the electron affinity of the anion. A more subtle point is that of whether electron affinity can be spectroscopically defined under these circumstances. This can be done if one looks at the vibrational states both of the anions and of the corresponding neutral species. The electron affinity is now defined from a specific line in the spectrum (see below).

One can compare this anion/neutral species technique with the previous cation techniques. Overall, one is talking about three different techniques (Table 17.1). The first technique was photoelectron spectroscopy, which involves energetically 'overshooting'. One photoexcites in the VUV region

Table 17.1. *Three different types of spectroscopy using photoelectrons*

Photoelectron spectroscopy molecular orbitals Turner, Kurbatov, Vilesov, Terenin	*PES*
X-ray photoelectron spectroscopy inner shell atomic orbitals Siegbahn	*ESCA/XPS*
Photo-ionization efficiency spectroscopy ionization energies Watanabe	*PIE*
Threshold photoelectron spectroscopy Peatman, Borne, Guyon	*TPES*
Zero kinetic energy spectroscopy ZEKE of Rydberg electrons ZEKE of positive ions Coincidence ZEKE spectroscopy Anion-ZEKE spectroscopy	*ZEKE*

with a very short wavelength and measures the energy of the resultant emitted electrons. This is the classic work that Turner and Terenin in the VUV[93,94,172] and Siegbahn in the XUV region[96] pioneered.

Photoelectron spectroscopy is very important for anions as a quick spectral survey. In an overshoot experiment one measures the electron affinity and analyses the electron energy, which is an important guide for setting the laser wavelengths, which are here hard to find.

Now one can scan the energy by increasing the photoexcitation energy and the electron current as the photoexcitation energy increases above the IP. At the IP one begins to measure a current which then increases to produce the Photo-ionization Efficiency (PIE) curve that Watanabe pioneered. For anions analogous experiments can be done.

The direct analogue is to measure the total current just as in a **Photo-ionization Efficiency** measurement (PIE), in which the total current increase is measured (Fig. 17.1). This was the photo-ionization efficiency for the cation and now the experiment is done in the same way, except it has the new name, **Photodetachment Spectroscopy** (PDS) and goes again as $E^{\ell+1/2}$. This was developed by a number of groups over many years, most prominently by Lineberger's group at Boulder[174]. Such a photodetachment spectrum is the analogue to PIE for measuring cation spectra. Again, it is the total current which is measured as the energy increases

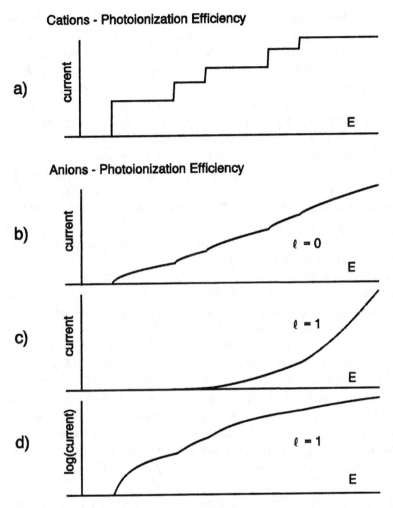

Fig. 17.1 The photo-ionization current versus energy. This illustrates the Wigner threshold laws. For cations one has the well-known staircase function (a). For anions s-waves ($\ell=0$) are shown in (b); p-waves ($\ell=1$) are hard to measure (c) since they have an asymptotic onset, which can be sharpened in a log plot (d). Most complex molecules will probably be unproblematic and have s-wave behaviour[173].

The current rises according to the Wigner threshold rules[173] as $E^{\ell+1/2}$, where E is the energy excess above the transition and $\ell=0,1$ etc. depending on whether the state of the outgoing electron is an s-wave, p-wave, etc. (Fig. 17.1). This in turn is fixed by the orbital in the molecule from which the electron is ejected and, hence, can become a signature of the orbital of origin.

Since in ZEKE spectroscopy a threshold is measured, it is found that $\ell = 0$ produces a $\sqrt{E}$ dependence that is observable at threshold. Higher values of ℓ are generally not observable at threshold, i.e. ZEKE spectroscopy in general will detect only s-waves, but this appears to be the dominant situation since complex molecules are expected to produce s-waves.

In ZEKE spectroscopy one examines only those states whose energy matches the photon energy. Under these conditions, the zero-energy electron is observed as a threshold signal and sharp peaks are observed. Any excitation away from these states creates hot electrons and, hence, will not contribute to the ZEKE signal. The ZEKE electron is unique also for anions in the sense that it produces sharp lines again. Now life is a little different for anions since the ZEKE states survive during their drawing out, which has been very useful for studying cations. One may have a dipole bound or image state in some cases, but their relative importance is theoretically still a matter of some debate.

One could perform a photoelectron-type coincidence measurement as well if one could go to higher laser repetition rates. The problems of anions and cations are quite different in theory and experiment and they lead to different results, which needs to be emphasized.

The easiest way to understand the overall scheme of how these methods compare is to put all three cases on a single diagram (Fig. 17.2). For cations, this corresponds to the right-hand half of this picture, but for anions, one must consider the left-hand half of this picture. Certain analogies can be found but the cases are not identical. I have drawn three similar potential curves, but they are, of course, different. The neutral species case would probably be the shallowest curve whereas the negative one is typically bound a little more and the cation case is probably the most strongly bound.

Let us now focus on the production of neutral species from anions. This technique is at present still much more difficult experimentally than is cation ZEKE spectroscopy because one photodetaches the anion.

The way you can do mass-selected neutral species spectroscopy, which appears to be a contradiction in terms, is to start with the anion at lower energy, M^-. This is then excited to form the M neutral species. You can see in Fig. 17.2 (right-hand side) the lines which would give the spectrum of the ground state of this neutral species obtained directly via the back door created by examining M^-. This will be particularly useful when dealing with species that are unstable as neutral species. They must, of course, be stable in the anion source, albeit as a mixture. As neutral species, they may have a very short lifetime, as in the case of transition states, some of which

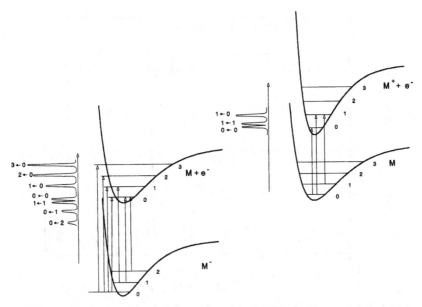

Fig. 17.2 An ionization scheme showing anion ZEKE spectroscopy (on the left-hand side), which displays lines for M^- and the mass-selected ground state M. The right-hand side displays the photo-ionization of neutral species to make the cationic ZEKE states.

survive less than 0.5 ps[175]. The anion, however, is required to survive at least for the flight time of a few microseconds between anion formation and photodetachment. Shorter times, however, could be achieved in a pump–probe configuration.

The spectrum produced in Fig. 17.2 (left-hand side) is analogous to a normal spectrum for neutral species ground-state spectroscopy. It has, however, a series of interesting new aspects not found in normal neutral species spectra. One interesting aspect involves the higher energy vibrational progressions. This might be of use as an efficient method for overtone spectroscopy of neutral species. Owing to selection rules forbidding $\Delta v \neq 1$ transitions, however, this is often an extremely difficult result to realize in normal ground-state spectroscopy. Recent work has shown that these higher overtones become very interesting to study. In the direct technique of ground-state excitation they are typically extremely faint. This is to be expected for a spectroscopy that is confined within this neutral species manifold. Selection rules are certainly more gracious when you start with M^- and go into the neutral system by removing an electron.

To illustrate anion ZEKE spectroscopy in a practical way, I have chosen

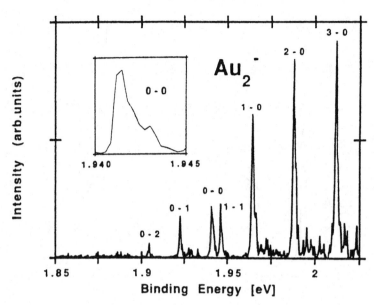

Fig. 17.3 The spectrum of the gold dimer. Note that the spectrum gives the vibra-ions both for Au_2 and for Au_2^{-157}.

Au_2 (Fig. 17.3), which illustrates an early case, namely, an experiment done by Kaldor *et al.* at Exxon[157]. This clearly shows a simple species, the gold dimer, which you cannot obtain from a bottle. For a typical jet with laser vaporization, a gold spectrum contains evidence of many clusters. With a mass spectrometer, one can select one particular negative ion cluster species. ZEKE detachment spectroscopy also permits one to focus on neutral Au_2 by setting the mass for the Au_2^- precursor. Note that, even though it detaches, it is still a ZEKE experiment, i.e., it does not measure the total current, but only the threshold electrons. It remains amplified with the ZEKE technique. This gives two spectra, i.e., both the anion and the neutral species data can be recorded. In fact, if the Au_2 neutral species spectrum were measured starting from the 0–0 origin of the Au_2^-, then, as the energy increases, the excitation spectrum would be measured. All the vibrations of the Au_2 neutral species are observed directly. Conversely, one can also see the vibrations of the Au_2^- negative ion with the corresponding transitions. The resolution in these early results was not terribly good, but they served to demonstrate very nicely the power of the method. An interesting side feature was that overtone intensities appeared strongly. This aspect suggests that these overtone intensities really can be quite interesting and useful for a new facile overtone spectroscopy. This indirect approach

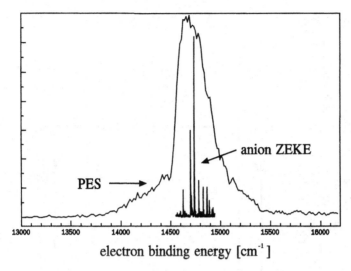

electron binding energy [cm^{-1}]

Fig. 17.4 Anion PES versus anion ZEKE spectra of OH^{-169}.

to ground-state systems should have applications far exceeding present examples.

Let me give you another example of seeing an anion photoelectron spectrum. In this case, I will take a negative ion, OH$^-$, as a simple system studied in some detail with prior techniques. Let us first look at the survey of a photoelectron spectrum as shown in Fig. 17.4. It is disappointing from the viewpoint of resolution but interesting in that only a single peak and nothing more can be observed. Nevertheless, one can identify the spectral region of interest. This is obtained from a mass-selected OH$^-$ in a negative beam, but only a single peak is found in PES, albeit for mass-selected OH$^-$. Photoelectron spectroscopy here measures the spectral range and the electron affinity.

One can measure the ZEKE spectrum on the same energy scale for this anion, as we have done recently and which is also shown in Fig. 17.4. You see that underneath this single photoelectron PES peak there is a good deal of fine structure. Now, of course, this must be taken as a textbook example. The OH$^-$ system has been studied in detail and reported in the literature[169]. Let us compare our results with what one would get from a standard photoelectron detachment yield spectrum. Figure 17.5(a) shows a detailed photoelectron detachment spectrum of OH$^-$ from Hotop *et al.*[174]. Indeed you see the hills arising from the square-root-law behaviour which you expect for s-waves in a total current measurement. From the breaks in successive

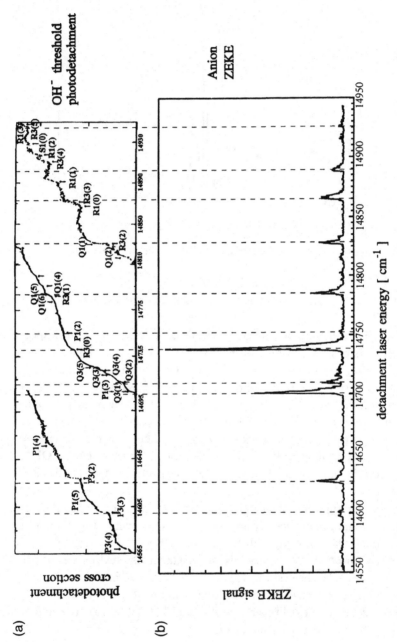

Fig. 17.5 The photodetachment yield versus the anion ZEKE of OH$^-$. (a) Shows the photodetachment spectrum of Schulz et al.[176]. (b) Shows the corresponding anion ZEKE spectrum[169].

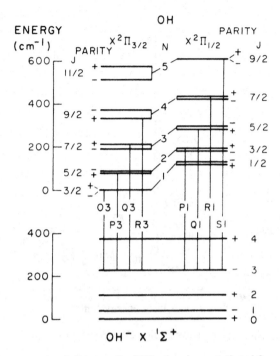

Fig. 17.6 The energy level diagram for OH$^-$ showing rotational fine structure, and Λ-doubling levels. The Λ-doubling splittings have been exaggerated by a factor of 50 so that they can be discerned. The threshold expected in photodetachment is labelled. See the text for explanation.

hills for photodetachment, small changes in shape, which distinguish various transitions, can be found. The level scheme explaining the nomenclature is shown in Fig. 17.6, e.g., R3(0) is the $\frac{3}{2}$ ground-state of OH coming from the zero state OH$^-$. This now represents detailed spectroscopic information that one obtains for this radical anion and radical neutral species system. If one expands the wavelength scale for the ZEKE transitions so that the ZEKE spectrum, now shown in Fig. 17.4, can be compared on the same scale with the photodetachment spectrum, then the spectrum appearing at the bottom of Fig. 17.5 is obtained. You can identify quite nicely the individual transitions as single lines from this ZEKE spectroscopy. This is a direct demonstration again of the difference between data from ZEKE spectroscopy at threshold and those from the cumulative spectroscopy of photodetachment. This is as one would expect for a mass-selected, anion–neutral species case. This demonstrates again that ZEKE spectroscopy also produces information for anions and neutral species, just as

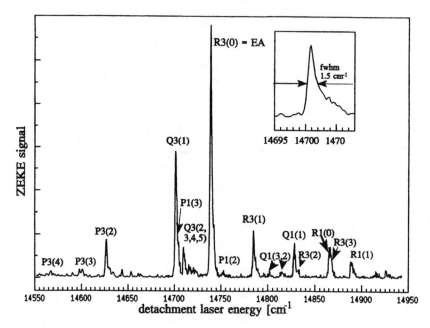

Fig. 17.7 Rotationally resolved anion ZEKE spectroscopy of OH^{-169}.

for the case of positive ions and hence it is a distinctive spectroscopy here as well.

Note that the electron affinity is not the energy needed to detach the electrons in PES, but it is, as expected in a spectroscopic experiment, the difference between two 0–0 states, i.e. it is the R3(0) line present in both spectra shown in Fig. 17.5. It is necessary to identify the spectroscopic transitions in detail, thereby to define the electron affinity.

The previous spectra serve as a comparison of photoelectron detachment spectroscopy and ZEKE spectroscopy for anions/neutral species. The OH spectrum is seen together with the OH$^-$ spectrum. The spectrum splits up into three components, with one component based on parity selection rules. One can identify all the various lines and compare them with the level scheme of Fig. 17.7.

The resolution one has up to now is slightly less than that for cation spectroscopy. These experiments are considerably more difficult to perform than they are for cations, but, nevertheless, a resolution of about 1–5 cm^{-1} can be obtained in this spectrum, which is quite respectable at this stage. It is at present the best spectral resolution for this anion/neutral species spectroscopy.

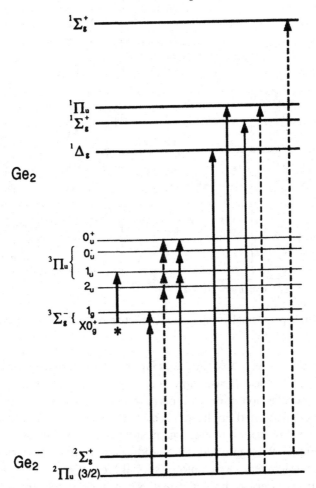

Fig. 17.8 One-electron photodetachment transitions between the two Ge^-_2 and six Ge_2 electronic states. The s-wave transitions are indicated by solid lines, p-wave ones by dashed lines[177].

Another example, similar to Au_2, but more recent, involves Ge_2 and Ge_2^{-}[177]. Here (Fig. 17.8), the level scheme for the six lowest-lying triplet and singlet states of the neutral Ge_2 is used. The s-wave transitions are observed directly and are shown with the solid lines – the dashed lines are p-wave transitions. This demonstrates a very important novel feature of anion spectroscopy, here shown in a nice example in which both singlet and triplet states of neutral species are observed directly in ZEKE spectroscopy. This has not yet been exploited. As can be seen in Fig. 17.1 the s-waves are detected directly in ZEKE spectroscopy, the p-waves are difficult to see

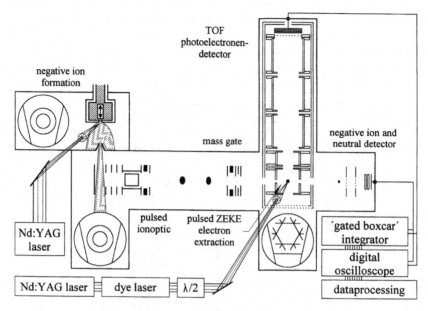

Fig. 17.9   The experimental set-up for mass-selective anion ZEKE spectroscopy[178].

because of the unfavourable threshold law, but both are seen in PES. This appears to be a special case, however, since most transitions of more complex molecules do not present this problem.

A more sophisticated application encompasses adducts such as those involving novel transition metal–organic intermediates. The experiment is shown in Fig. 17.9. Negative ions are produced in front of a nozzle source by electron attachment. We employ a special kind of electron production from the strong laser irradiation of metals causing electrons to be emitted and, thereby, obtain a highly pulsed current. This results in one pulse of electrons giving kilo-amperes for efficient attachment[179]. After production of the anion mixture, the ions pass through a skimmer to the pulsed ion optics of a mass spectrometer. The beam is aligned in a right-angle configuration and the ions are drawn out in a direction perpendicular to the cluster beam. At a point down the drift tube, the light pulse is added to obtain ZEKE excitation. The ZEKE electrons are drawn out again in a perpendicular direction, whereas the neutral species can be monitored with an in-line detector. To illustrate how well this mass-selective laser excitation works, a reflecting grid was incorporated to separate neutral species from anions. This apparatus is described in more detail in Fig. 17.10.

At the beginning, shown in Fig. 17.10(a), the reflector and laser are switched off. This is the simplest case of all. The anion spectrum is produced from the interaction of the laser-illuminated metal tip with the ethylene molecules seeded in the jet. Here one sees the total anion mass spectrum including the iron–ethylene adducts in the spectrum. The anions being produced contain, among other species, various complexes of iron and ethylene.

Laser desorption of iron was used in the cluster source, since we want to study how ethylene is tied to iron in a complex. This might be analogous to the bonding which is involved when ethylene is bound to a metal surface, as in certain catalytic mechanisms. In these experiments, ZEKE spectroscopy is used to study metal complex adducts. In this first case, the beam is coming all the way from the source to the end of the linear drift region where the ions are detected by a multichannel plate (MCP) detector. The reflector is switched off. Since the anion cluster beam enters the source of the mass spectrometer at right angles, only ions from the primary beam are detected, thus preventing neutral species created in the cluster beam from reaching the detector. This means that an ordinary anion mass spectrum is measured using a **Linear Time-of-Flight (LINTOF)** mass spectrometer. As before with cation mass spectra, you now have measured anion mass spectra.

For a second set of conditions, the same signal is measured except that now the reflector is turned on. No ion arrives at the final MCP detector since every ion gets turned around by the reflecting grid. Owing to this reflection grid, there is no straight-through signal after the LINTOF portion, either from ions or from neutral species. Hence the second scan in Fig. 17.10 is observed, i.e. no signal is produced. This is necessary, because you have to show that there is *no* collision-induced detachment or decay of metastable anions.

Consider now the third experiment, at which point the reflector is still switched on, but the detachment laser is switched on as well. The laser timing is set to the arrival time of an anion of a given mass at the laser focus. The laser picks off the electron from one species, which is pre-selected. Since the neutral species thus formed is not turned around by the reflector, a new peak in the MCP appears at the end of the LINTOF. The appearance of a single neutral species mass is the proof of mass-selective photodetachment. If, in addition, you select the proper wavelength, the electron will be split off and spectroscopic information about this mass-selected neutral species can be obtained.

Thus, one can produce a complete anion mass spectrum and pick out one species for a PES or a ZEKE spectrum. This is shown here for the case of

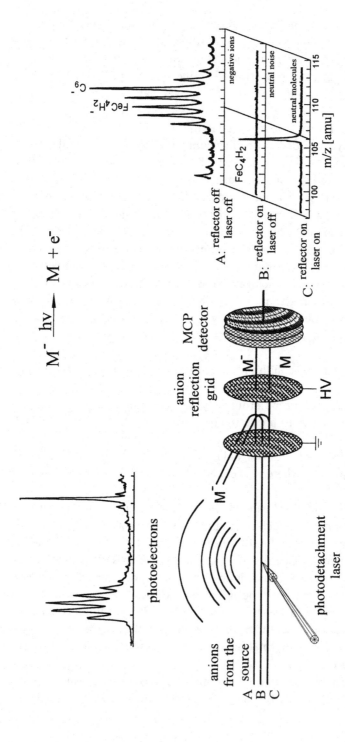

Fig. 17.10 Mass-selective photodetachment. Note that in (a) one just has a standard anion mass spectrum of the adduct species. In (b) one turns on the reflector. This deflects all ions produced in (a). In (c) one turns on the laser to produce specific neutral species by photodetachment. These neutral species are now not deflected in (b) but pass to the detector to give the derived neutral species signal[180].

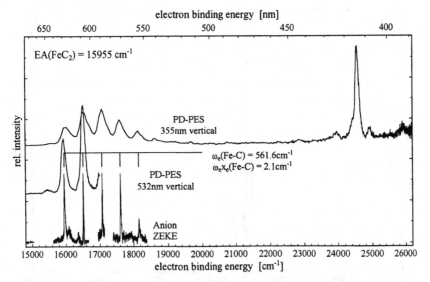

electron binding energy [nm]

650   600   550   500   450   400

EA(FeC$_2$) = 15955 cm^{-1}

rel. intensity

PD-PES
355nm vertical

ω_e(Fe-C) = 561.6cm^{-1}
$\omega_e x_e$(Fe-C) = 2.1cm^{-1}

PD-PES
532nm vertical

Anion
ZEKE

15000 16000 17000 18000 19000 20000 21000 22000 23000 24000 25000 26000

electron binding energy [cm^{-1}]

Fig. 17.11 The photodetachment photoelectron spectrum and anion ZEKE spectrum of ^{56}Fe^{12}C$_2^{-}$[170,178].

an organo-metallic, short-lived intermediate species spectrum, measured in its ground neutral state.

The final spectrum now can be seen for FeC$_2$ as a neutral ground state spectrum in Fig. 17.11. This again starts with FeC$_2^-$ being picked from the anion mass spectrum. The ZEKE spectrum is observed as high-resolution lines. The photoelectron spectrum which appears as broad lines provides a rough approximation of the Fe—C stretch features at 550 cm^{-1}. The progression for the Fe—C stretch mode can be seen readily and from this progression it can be concluded that the iron must sit collinearly with respect to the C—C structure. This is an important point indicating the bonding character in the organo-metallic intermediate.

Another example of a spectrum is obtained for S$^-$ (Fig. 17.12), which is a simpler case in which a fairly broad photoelectron spectrum is observed. If we do this in ZEKE resolution, it would be somewhat compressed on this scale as before (Fig. 17.12), although the resolution seen in this particular case remains clear. We can also show what the ZEKE peak looks like compared with the photodetachment peak. This is shown in the lower half of Fig. 17.12. The ZEKE resolution is 1.6 cm^{-1} and the total photodetachment spectrum increases like the old staircase function, having the expected $\sqrt{E}$ rounding effect due to the s-wave detachment. This photodetachment is difficult to interpret, as can be seen in the plot. The comparison with the

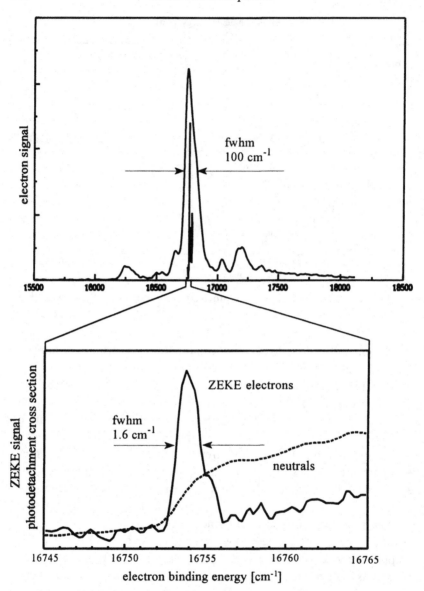

Fig. 17.12 Anion PES and anion ZEKE spectra of S^-: the lower spectrum is the expanded ZEKE spectrum together with the total current, as in photoelectron detachment experiments[90].

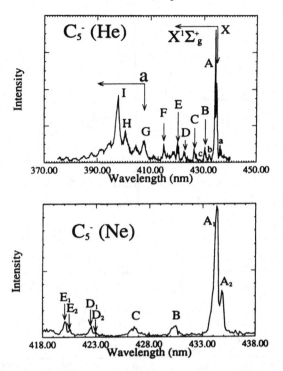

Fig. 17.13 The anion ZEKE spectrum of carbon clusters in different expansion gases.

ZEKE peak is shown in the spectrum, but the ZEKE peak gives a clearer indication of the proper energy. This shows the difference between photo-detachment spectroscopy and ZEKE spectroscopy for these two cases. The comparison with the top spectrum in Fig. 17.2 shows the increased resolution of ZEKE by stretching the wavelength scale. Some more complicated systems involve, for example, clusters such as C_5^- and neutral C_5 (Fig. 17.13), which are shown for different expansion conditions[162].

Higher level clusters of anions have been investigated with photoelectron spectroscopy by Chesnovsky, Neumark, Smalley etc.[181] Arnold *et al.*[181b] looked at the case of photoelectron spectra from halogens solvated by CO_2 clusters (Fig. 17.14). The dotted curves are the Franck–Condon simulations, which fit the data quite well. This shows that here the Franck–Condon principle works quite well even in ZEKE spectra. Such a correspondence is not always seen for ZEKE experiments, in which there are often interesting additional bands. It is interesting because one is interested in solvation in an open-shell system such as iodine, here, $I \cdot CO_2$. First,

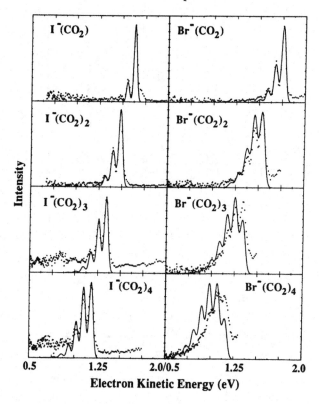

Fig. 17.14 Photoelectron spectra versus Franck–Condon simulations[181b].

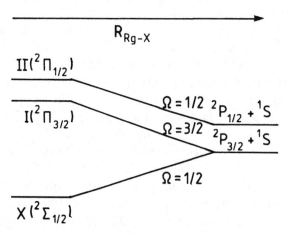

Fig. 17.15 The correlation diagram for the two atomic states of iodine forming the three states of the $I \cdot CO_2$ cluster[182].

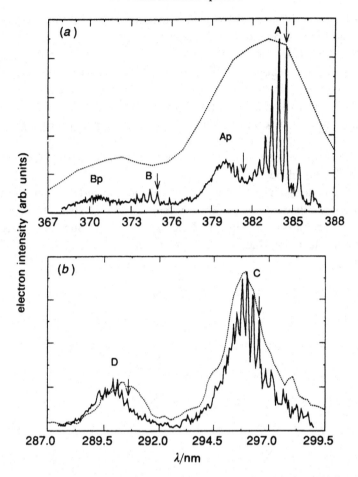

Fig. 17.16 The threshold photodetachment spectrum of $I^- \cdot CO_2$ (solid lines) and the photoelectron spectrum of $I^- \cdot CO_2$ (dotted lines): (a) the $I(^2P_{3/2}) \cdot CO_2$ band and (b) the $I(^2P_{1/2})$ band. The arrows indicate the band origins[182].

one has to remind oneself that iodine exists in two electronic states, the $\frac{3}{2}$ and $\frac{1}{2}$ states. If you draw the correlation diagram (Fig. 17.15) for the complex, the $\frac{3}{2}$ state splits into two states and, together with the $\frac{1}{2}$ state (Fig. 17.15), forms three states in the cluster. All three levels are seen in Fig. 17.16 as expected. The resolution is of the order of 8 cm^{-1}. Notice again the broad photoelectron spectrum in this spectrum at the back (dotted lines) from the work of Arnold *et al.*[181b] This gives you a very rough, but important, indication of the energy at which the spectrum is to be found. The ZEKE spectrum produces many lines in very nice progressions, which is interpreted as a linear C—I frequency, as in a diatomic system. Thus, this progression is interpreted as a T-shape complex between I and CO_2 and appears to indicate that the anion has a stronger bond than the neutral species and that this bond is shorter by some 0.22 Å due to the CO_2 quadrupole interaction with the polarizability of the I atom. There are two other prominent methods of gaining such interatomic potentials and interaction energies. For this case in particular, the scattering experiments by Lee's group provided appropriate information[183–187]. This information also has been found quite extensively by emission spectroscopy by Golde and Thrush[188] at Cambridge.

Lineberger is also investigating systems like iodine and bromine atoms as they become solvated in the overall system[167]. These systems are excellent candidates for mass selectivity and ZEKE spectroscopy. Such mass selectivity allows one to produce ZEKE spectra of the neutral species in the system.

18

Short-lived states

One of the most fascinating aspects of ZEKE spectroscopy is its application to short time scales. This leads to new applications in the direction of femtosecond excitation; the field called femtochemistry has opened up new perspectives in ZEKE spectroscopy. Already, some such applications have been published[189–191].

Some degree of caution is, however, required. The shorter the excitation time the larger the energy uncertainty, so that single-state population will be all but impossible for femtosecond excitation. Secondly, femtosecond lasers, though apparently operating at reasonable energy, deal with enormous intensities, which are then reflected in a multiphoton excitation scheme. Thirdly, femtosecond timing is so short that one can consider the nuclei to be nearly frozen on this time scale, i.e. all vibrational motion starts only after the laser pulse is over. This means that all molecular energy transfer processes, such as those leading to energy randomization such as intramolecular vibrational relaxation (IVR), start much later. A typical timescale for IVR is $\tau \simeq 1$ ps[192,193]. Reaction times themselves can be as slow as some 10 ns, if randomization is a prior process. To first order one would expect that any excitation with a pulse faster than 10 ns would then allow one to see this 10 ns decay. This is true for small molecules. For large systems this equivalence breaks down[194]. Prompt excitation on a femtosecond timescale leads to a chemical reaction totally different from that obtained with nanosecond excitation. The former process, carried out on frozen nuclei, leads to a strongly localized chemical reaction, whereas nanosecond pulses lead to a clear statistical process. Hence excitation below the IVR timescale leads to different chemistry for very large molecules. Such differences would be strongly reflected in the ZEKE states populated by the different fast or slow pulsed lasers. ZEKE spectroscopy of these short-lived chemical intermediates would be an important new area,

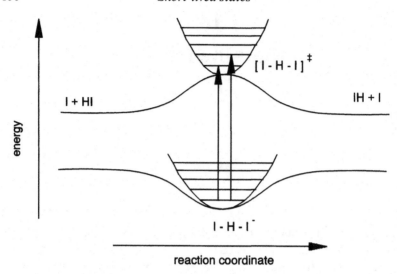

Fig. 18.1 ZEKE spectroscopy of the transition state (cf. Neumark[175]).

rendering the technique useful for the study of new reactive intermediates in chemical reactions.

18.1 Transition states of chemical reactions

One of the most fascinating applications is derived from the absolute rate theory of reaction rates of Polanyi, Eyring and Wigner (cf. Ref. 195). This theory postulates a critical surface crossing at the transition state. This transition state has an internal translation along the reaction coordinate and hence has been deemed to be invisible by spectroscopic means. Neumark demonstrated that this transition state can be seen directly in ZEKE spectroscopy by starting with the anion.

The example chosen here is the transition state ($\neq$) for the reaction between I and HI:

$$I+HI \rightleftarrows [I\!\!-\!\!H\!\!-\!\!I]^{\neq} = \rightleftarrows IH+I$$

This state can be produced directly by first making the stable $[I\!\!-\!\!H\!\!-\!\!I]^-$ and then doing ZEKE spectroscopy on the species, pumping the transition state $[I\!\!-\!\!H\!\!-\!\!I]^+$ directly in its orthogonal vibrations (Fig. 18.1).

A ZEKE spectrum recorded by Neumark in Berkeley[175] is shown in Fig. 18.2 for the famous transition state of IHI. He started again with the IHI$^-$ anion, picked off the electron in a ZEKE mode and, in this way, obtained the ZEKE spectrum of the transition state directly as a spectroscopically

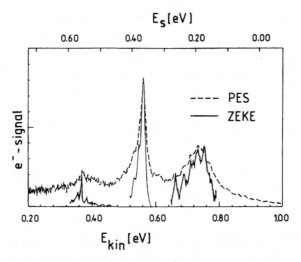

Fig. 18.2 The ZEKE spectrum of the transition state of a chemical reaction [I—H—I]. The dashed line is the PES spectrum[175].

accessible intermediate. This state has an estimated lifetime of at least 180 fs. Even something as fragile and transitory as an activated complex can be readily investigated by ZEKE spectroscopy. This represents one of the first direct measurements of the spectrum of a transition state in chemical reactions.

19

Applications – state selection

Obviously, ZEKE spectroscopy, particularly as a coincidence technique, is also a state selector (Table 19.1). In a coincidence experiment, the arrival time of the ZEKE electron signals the energy for the corresponding ion. Hence the ion in coincidence is state-selected. The direct measurement of the state-selected ion at the ZEKE level allows one to study the energy dependence of rates at various levels. This would provide highly accurate energy definitions for precise studies of unimolecular reactions. Similarly, one can look at ion–molecule reactions in which one of the partners is state-selected by this method softly. One could look at intermediates in chemical reactions and identify the structure of the intermediate involved. In multi-photon mass spectroscopy (Table 19.2) timescale measurements are of interest as well. These measurements lend themselves to measurement on

Table 19.1. *Applications of ZEKE spectroscopy*

Unimolecular reactions
Rate constants at defined energy can be varied in energy to obtain k(E)
Coincidence experiments with electrons and ions
PIPECO → ZEKE
Limits: coupling of soft modes and very large molecules

Ion–molecule reactions
State selection of reactants according to ZEKE state's translational energy
internal energy–state selection

Intermediates in chemical reactions
$$A+B \rightarrow [AB]^* = \rightarrow C+D$$
$$\uparrow$$
$$[AB]^-$$

Electron transfer and delayed ionization pump–probe experiments

---

Table 19.2. *Multiphoton excitation of ions in MS*

MUPI-MS 20 Hz to 10 kHz
Multiphoton excitation of ions in MS
Multiphoton ionization leads to dissociation of ions
Pump–probe – interrupted kinetics
fs to ns
Kinetics measured in the acceleration region
ns to μs
Kinetics measured in the drift region
μs to ms
Kinetics measured in ICR
ms to s

timescales as short as femtoseconds and have been applied here[191,196]. The important point is that all these techniques can now readily be extended to include the resolution of ZEKE spectroscopy, not only for cations but also for neutral species.

20

Channel interactions and selection rules

Now let us discuss Franck–Condon calculations, selection rules and ZEKE propensities due to channel interactions. Channel interactions refer to the coupling of Rydberg series, each of which is attached to a particular rotational or vibrational state of the ion in Hund's case (d) or the 'inverse Born–Oppenheimer' limit[117]. One fact that makes this important is the strong interaction between these Rydberg series, which permits entering any one of these series and arriving in another series. This was first shown for atomic systems by Mahon *et al.*[112] and Bordas *et al.*[114] and was described well for molecular systems by Berry and Nielsen[197]. Molecular examples have been given by Grant *et al.*[198] for NO_2, by Gerber and Bühler[199] for Na_2 and by Yeretzian *et al.*[200] for Ag_2. The field in relation to ZEKE has been reviewed by Merkt and Softley[121].

The term channel interaction will be employed rather than auto-ionization in order to develop our understanding of selection rules. What one sees are very strong perturbations of line intensities and line profiles in ZEKE spectra. Let me illustrate this in Fig. 20.1, which is a very simple spectrum of the silver dimer. This is a cation ZEKE spectrum, from resonant intermediate-state excitation of the silver dimer. The vibrational intensities for the various kinds of vibrations which are measured all seem to favour the zero vibrational state in the ion. One can calculate the corresponding Franck–Condon factors for these transitions starting from different intermediate-state levels with good accuracy, but these show a rather different set of propensities (Fig. 9.10). There seems to be a clear discrepancy between the observed ZEKE spectrum and the Franck–Condon intensities. This is not always so. In photoelectron spectroscopy, Franck–Condon calculations and the experiment usually agree[73]. For polyatomic systems like *p*-difluorobenzene, one employs these propensities as an aid to assignments. For silver it does not work at all. The answer may be found, at least

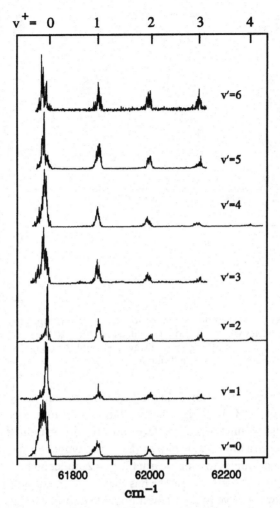

Fig. 20.1 Delayed pulsed field ionization of the silver dimer. Ionization starts from various vibrations ν' of the intermediate state B $^1\Pi_u$ to the various ν^+ of the ion.

for this particular system, in channel interactions. These are unique to ZEKE and appear to show up particularly for heavy diatomic molecules. They are of particular importance for a warm jet, in which a substantial rotational excitation is present.

When one talks about ZEKE excitation in a straightforward and direct way, one refers to ionization from just below the ionization potential, whereby one reaches a particular ZEKE level developed from its high-n Rydberg state. This is the customary method of ZEKE spectroscopy.

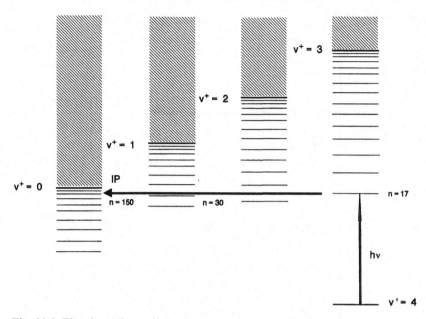

Fig. 20.2 The channel coupling scheme for Ag_2, showing the preference for the $v^+=0$ state no matter which manifold is excited according to the selection rules. The strong coupling among Rydberg states always seeks out the highest n, even if it is in a different stack.

However, if one starts from a higher vibrational state, say $v'=4$ in the REMPI spectrum of Ag_2 (Fig. 20.2), then one will pump a progression of Rydberg levels leading to the $v^+=3$ level most readily. Among the states one will pump maybe a lower value of n of the $v^+=3$ Rydberg progression. Now this level in turn couples to a higher n at the $v^+=2$ ladder and so on until near the $n=150$ state of $v^+=0$, where one has the direct transition to $v^+=0$. The system will always drive to the highest possible n since this has the highest state density. Hence it is entropy-driven. This is, of course, subject to the fact that this coupling is possible all the way to the highest n. For poorly coupled manifolds, such as is expected for wide level spacing, this migration may be inhibited, so the low-n state populated will usually decay rapidly and not contribute to the ZEKE spectrum until higher energies. Hence this appears like $n=150$ of the direct transition but coupled from a different state v^+. The coupling starts from $n=17$ and moves by channel coupling to $n=150$. Here it forms a high-ℓ ZEKE state with a large m_ℓ and hence larger $(2m_\ell+1)$ phase space. This is an example of channel coupling and you might consider it rather as a back-door way of obtaining the ZEKE state with only an apparent circumvention of selection rules. There are many examples of this type.

It is an easy way to avoid a Franck–Condon gap in ZEKE spectroscopy. This channel coupling is a real and prevalent mechanism[121], although its importance is quite different for differing molecular systems. This mechanism has the virtue that you have the ability to induce transitions that you would not see otherwise. Here Franck–Condon propensities induce apparent violations of the Franck–Condon principle and also apparent violations of selection rules in ZEKE spectra. The bad news is that your intensities are not readily interpretable by standard two-state theory but require a more sophisticated treatment. The standard model employs Franck–Condon selection rules and has been shown by Wang and McKoy[73] to work extremely well for many ZEKE spectra. When channel interactions become prominent, one sees intensities even though they are forbidden by selection rules by many orders of magnitude[201]. This is the direct result of channel coupling. Dickinson *et al.* have shown[146] recently that even these ZEKE intensities can now be calculated employing MQDT theory[202]. Here one makes use of the fact that the ZEKE electron while far away is uncoupled from the ionic core whereas close in it is coupled to the ionic core. The intensities can be understood quantitatively in terms of this frame transformation. The good news is that you see many more types of spectroscopic transitions and hence in practice have access to virtually all energetically feasible energy levels. This could, in principle, also be used to produce entirely new spectra that normally are obscured by the Franck–Condon principle.

The most basic example can be seen in terms of traditional auto-ionization. Let us review auto-ionization. The simplest, most basic schematic description of auto-ionization is shown in Fig. 20.3. One can have direct ionization involving one set of levels, while one has other levels nearby that are coupled to the first set of levels by various kinds of auto-ionization processes. If the Rydberg states of B are above the IP of A, coupling to the continuum states of A occurs and direct auto-ionization is observed. If the Rydberg states of B couple to bound Rydberg states of A the ZEKE intensity of A is modified.

Several theories have been suggested that embody the typical propensity rules (not selection rules). Berry has done detailed work on explaining these propensity rules[203]. Since these mechanisms will recur over and over again in ZEKE spectroscopy one wants to view these coupling processes generically rather than limiting oneself to definitions based on classical auto-ionization perturbation. Let us mention two aspects of auto-ionization.

1. There can be more than one stack of levels involved in auto-ionization. This coupling between many stacks involved in auto-ionization will here

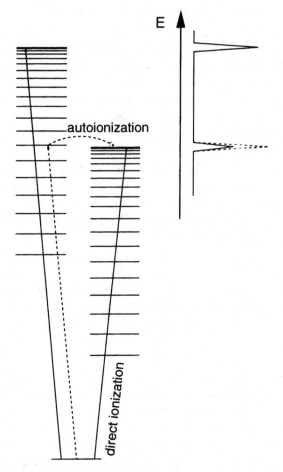

Fig. 20.3 The auto-ionization mechanism coupling two Rydberg stacks. The interference between the levels modifies the ZEKE intensities.

be replaced by the more general concept of channel coupling (Fig. 20.4),
which is a more current term. Channel coupling reflects the enhancement
and the decrement produced by the auto-ionization process in a more
general fashion. It can involve few stacks or many stacks. The whole as
well as the enhancement of a spectrum in a multi-state system is considered. The bad news associated with auto-ionization is that it falsifies the
intensities. I shall show such an effect in a spectrum.

2. The good news, however, was something quite surprising: via channel
   coupling, one can sometimes stimulate the population of new levels or,
   simply stated, circumvent selection rules, which would otherwise prevent

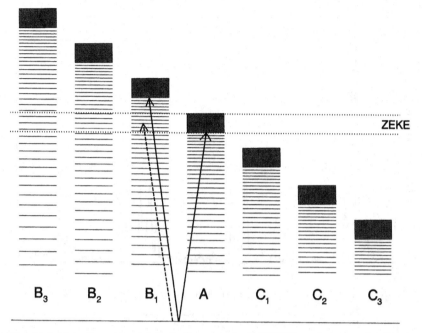

Fig. 20.4 Coupling of multiple Rydberg stacks in ZEKE spectroscopy leading to channel interactions including vibrations and rotations (cf. Merkt *et al.*[121]).

us from observing any transitions to certain quasi-forbidden states. Collectively, our recent experiments indicate that one can circumvent most of the selection rules, in one way or another, and thus use channel coupling to our advantage. This is not to say that Franck–Condon factors are never operative; they certainly are in the original optical transition, but they may or may not be dominant in the final spectrum. They do, however, sometimes determine propensities. This would be particularly so for complex molecules. In such a case, they are useful in a tandem assignment of the ionic states from the originating states. One can abuse direct transitions, so to speak, by adding channel-coupling-type transitions to allow access to states that one otherwise could not see. There is, therefore, a simple aspect and a complex aspect to the ionization process.

To show you the falsification of the intensity due to auto-ionization, consider a case in which you know the answer. The example is hydrogen (Fig. 20.5), for which the experimental spectrum was observed by Peatman[204] with very large false intensities. If one looks at the Rydberg series for hydro-

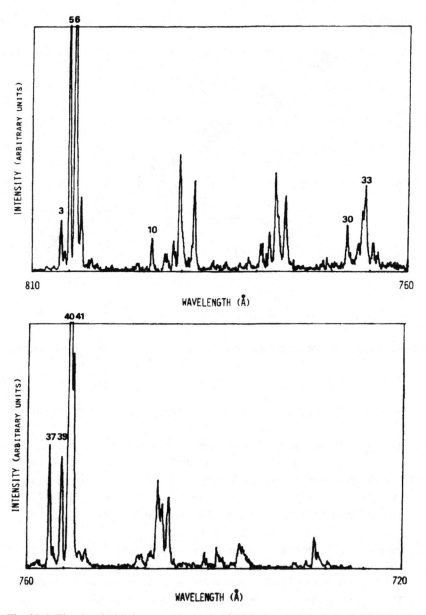

Fig. 20.5 The threshold electron spectrum of hydrogen (resolution: about 3 meV (25 cm^{-1})). Note the strong intensity perturbation due to interfering auto-ionization resonances[204].

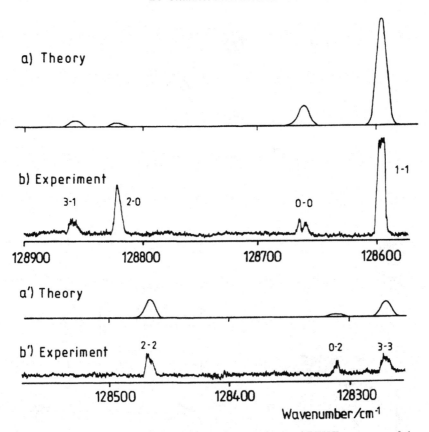

Fig. 20.6 Hydrogen. The predicted versus the experimental ZEKE spectrum of the $X\ ^2\Sigma_g^+(v^+=2)\leftarrow X\ ^1\Sigma_g^+(v=0)$ transition. This shows the intensity changes due to auto-ionization coupling of Rydberg levels[121].

gen, one can discover the auto-ionization states which are responsible for these difficulties and even identify them. When one compares the theoretical spectrum for hydrogen (top of Fig. 20.6) with the experimental curve, the same results are sometimes found; but, in some bands, very strong anomalies are observed, e.g. the 2–0 transition.

If one looks at these intensity anomalies more carefully for the cases of hydrogen and nitrogen (nitrogen will not be discussed here) one finds many resonances in the scan that display these anomalies. Since the hydrogen spectrum is understood well, these anomalies can be readily assigned to perturbations from other known Rydberg states. This can be seen in Fig. 20.7. If one is going to the ionization potential in Fig. 20.7, one can arbitrarily divide the region below the IP into a region (a) just below the IP and

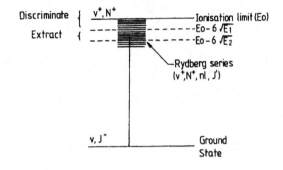

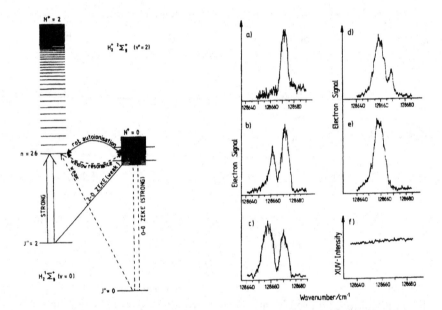

Fig. 20.7 The hydrogen window resonance ($N^+=0\leftarrow J''=0$). Note the changes in intensity (a)–(f) as the spectrum is scanned through the window resonance shown on the bottom left-hand side[121].

a second one (b) a little lower. The region (a) is involved directly, due to there being a spoiling field to discriminate against prompt electrons. The region (b) is field-ionized and leads to the ZEKE signal. There are at least two regions that one can consider. A strong 0–0 transition is observed for (a) with a width corresponding to the depth given by the pulsed ionizing field. As one decreases the energy in case (b), a second peak apparently appears, but it is really the same peak with a window in the middle due to an interaction with a new manifold. Here the $J''=2$ state possesses a transition from $n=26$ of its own Rydberg series to the $N^+=2$ ion state which is coupled to the original IP. Since this is a strong transition $n=26$ is short-lived; it is a drain on the original ZEKE state population and, hence, a 'window resonance'. As the energy due to an increase in the spoiling field decreases the window eventually disappears when a mismatch to the $n=26$ state occurs. One has a typical rotational auto-ionization, from which one would expect to produce mixing of these states, but, of course, this also produces these window resonances. In other words, one expects to get contributions that result in positive and negative intensity changes to the direct transitions and, in this way, a weak transition of two of these states can be seen to interfere. You can actually identify which transitions would interfere most strongly. You look at various resonances on going from (a), (b), (c), (d) and (e) and you notice a very nice ZEKE transition. Then you change the energy by going down into the Rydberg manifold. By further dipping into the lower energies, you see the window leading to a splitting in the peak and then a second peak appears. The window resonance that detracts from this state appears squarely in the middle, which permits the assignment of this resonance. Hydrogen is an ideal case because you can really show which are the actual transitions and which are involved when you have both positive and negative influences on the ZEKE spectrum due to these interferences.

In the case of hydrogen, of course, life is particularly simple and predictable since it involves a two-step system between two states. In actuality, however, life is a great deal more complex than this, even for nitrogen. Many kinds of possible transitions for this type of interaction can be drawn with a whole series of Rydberg manifolds since there is a manifold associated with each rotational and vibrational state of the ion. This, of course, leads to the complications and the new possible perturbations involved in determining ZEKE intensities[200,205].

In Fig. 20.4, the ionization potential lies at the edge for the ZEKE excitation in the A channel. The normal ZEKE states are operative with high-n excitation and each one of these states is long-lived near the ionization continuum. In the background, however, there are other continua, each foreshadowed in their own Rydberg series leading to their own state convergence. In an isoenergetic sense, each of these series couples and either steals or borrows the intensity to cause changes in these original ZEKE bands.

Let us examine the coupling from the left. B_1 provides resonances at lower n Rydberg states but B_2 has resonances at a still lower n. This pattern continues for all B values. Since the lifetimes of the lower n Rydberg states are shorter and more widely spaced this coupling for ZEKE spectroscopy is less important for B_2 than it is for B_1 going to A. This means that coupling is most effective for states for which the decrement to the final states is not too large and for which n increases. On going to the right-hand side from A, only those continua which can lead to shape resonances appear[206,207]. Hence, the main interaction will be with a few state Rydberg series on the left-hand side of A, which form in the main those series leading to near neighbouring states.

The series to the left of A will contribute to the intensities, but generally only weakly unless these are preferably pumped by selection rules. This can provide a channel for optical absorption. If, for example, B_1 has a better oscillator strength than A for absorption, then one can have absorption via B_1 and coupling into A, thereby producing a ZEKE state at A as if A had been excited directly. This leads to an effective circumvention of the selection rules, even though they hold for initial absorption, but in the B_1 ladder. Hence the channels to the left of A are effective channels for pumping when the A channel is blocked and thus are apparently circumventing selection rules.

The mechanism is the population by absorption of a given n, for example in the B_1 ladder. This provides coupling to higher n in the A ladder. Higher n values have longer lifetimes and larger phase space, and therefore dominate the ZEKE spectrum, causing the spectrum to contain the eigenstate signature of A even though excitation is via B_1. Generalizing, one can say that a lower n state of an allowed ladder is always pumped if the direct excitation is forbidden, but the spectral signature will be that of the forbidden ladder. This is an important point, since it allows us practically to observe many, if not most, energetically allowed states via ZEKE spectroscopy, even if they are forbidden by conventional spectroscopic rules. It should be remembered that the conventional application of such rules only refers to direct transitions and not to channel couplings. The presence of channel couplings only appears to circumvent these selection rules. In fact, they continue to hold separately for each individual stack of transitions.

Six aspects need to be remembered.

1. This coupling between Rydberg ladders only works effectively if the Rydberg states in the various ladders are closely spaced energetically. This is the case when a dense manifold of rotational and vibrational states exists, each with its own Rydberg series, coupling together. A

heavy-atom system, such as Ag_2, would be such an example. In special cases Guyon *et al.* have shown that pre-dissociation continua can couple these stacks[208].

2. The various Rydberg ladder coupling schemes work as follows. For the highest intensity signal, excitation typically starts at $v'=1$ in the originating intermediate state (assuming $\Delta v=0$ selection for simplicity). Since one can start at $v'=1$, one can reach the IP ($v^+=0$) at slightly lower energies and, hence, longer wavelengths. Nearly degenerate with the IP is, say, the $n=50$ state of the B_1 Rydberg ladder, which through optical transitions in this ladder leads to $v^+=1$ (Fig. 20.4 and Fig. 20.2). This n state couples to the IP Rydberg manifold at $v^+=0$, i.e. the two manifolds, B_1 and A, couple such that B_1 is in $n=30$ and A is in a ZEKE state with a lifetime 30–200 times longer than that of the $n=30$ state, i.e. all the population is in the B manifold, just as if it had been produced there directly. Hence, this produces the ZEKE state via the 'back door' and at the lower energy of the IP. The B_2 stack is less suitable for this mechanism since the density of B_2 Rydberg states in the neighbourhood of the IP is even less than it was for the B_1 ladder. Of course, the closer the levels in the ions are spaced, the more similar will be the density of Rydberg states between B_2 and B_1.

3. Rotational coupling schemes. Consider the initial excitation: as for the case of hydrogen (Fig. 20.7) the $J=2$ state could be the originating state, which here is taken to correlate with the $N^+=2$ state of the ion. Again, the back-door transition to the IP ($v^+=0$) becomes possible at an energy to the red of the 0–0 transition, but only from the Rydberg ladder of the $N^+=2$ rotational state. This is repeated for all rotational ladders that are accessible. Access to rotational ladders is determined by the distribution of originating rotational states for the transition. This means that, in this extreme, the onset of the IP will be red-shifted by as many originating states, J, as are thermally excited. Hence, all these ionization potentials are red-degraded by the rotational envelope. The higher the mass the smaller the rotational spacing and the better the near degeneracy of the ladders that is required in order for the back-door mechanism to work.

4. Consider two rotational states, one at $N^+=0$ and the other at $N^+=20$, each with their own closely spaced Rydberg series coupling to the IP ($v^+=0$). In order for the ZEKE state at the IP to have a long lifetime, it must be in a high-ℓ state; (e.g.) $\ell=20$–30 would fulfil the optical requirements within the B ladder. Photoexcitation in the $N^+=0$ manifold will be near $\ell=1$, from which the system has first to migrate to a higher ℓ state

in order to survive as a ZEKE state. The $N^+=20$, $n=50$ state offers an alternative back-door mechanism for entering this ZEKE state. Since it is in a high rotational state, a transition into the $\ell=20$, $n=150$ ZEKE state may well be facilitated. Hence, the mechanism via the $N^+=20$ ladder at $J'=20$, $n=50$ may actually be more efficient than that producing the $n=150$, $\ell=1$ state from the $v'=0$, $J'=0$ ground state.

5. Consider competing mechanisms on the ladder starting at $v^+=1$ and $J^+=20$ compared with the direct A ladder. By starting at $v^+=1$ a corresponding red shift in the ionization to $v^+=0$ is produced, which efficiently utilizes a back-door mechanism via rotational excitation to $N^+=20$ and $v^+=1$. As the energy increases to reach the direct 0–0 transition and hence the IP on the A ladder, this energy will also roughly suffice to reach $v^+=1$ from $v'=1$ by a direct mechanism and, in turn, by an indirect mechanism from the back door of the B_2 ladder. This will then interfere with the direct mechanism. The transition for $N^+=20$ in the B_2 ladder will, however, be the stronger transition.

6. A further case would occur if the levels were split by a small field. This would provide level matching and improved coupling between Rydberg stacks (see below).

The conclusion for the special case of dense Rydberg manifolds is that almost all states are populated via the 'channel-coupling' mechanism, i.e. by Rydberg ladders leading to higher states degrading to long-lived ZEKE states at lower v^+ and higher n. In the extreme, this means that no matter how you populate the levels v^+, starting at any level $v'=0, 1, 2, 3, 4, 5, 6$, the $v^+=0$ is always the strongest transition here. This is clearly seen for Ag_2. For lighter systems the effect will not be that dramatic. Also, it must be remembered that channel coupling to reach the red-degraded states is mandated by the selection rules constraining optical transitions to remain within the Rydberg ladder.

It also must be remembered that all vibrations in the ion are red-degraded by the width of the rotational distribution that is populated thermally in the originating state. Hence this problem disappears as $T \rightarrow 0$ K. These effects are clearly observed for Ag_2. It is an apparent (but not real) failure of the Franck–Condon principle, if the interactions with other ladders are considered. The blue edge of all transitions represents the true spectroscopic value for the transition (at $T=0$ K).

This effect is operative uniquely for a high density of Rydberg levels, i.e. heavy molecules at non-zero temperatures. The situation is often different for light molecules. The Franck–Condon calculations work well, as has

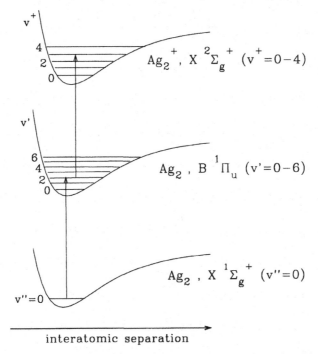

interatomic separation

Fig. 20.8 Resonant two-photon/two-colour excitation of Ag_2^+.

been shown for hydrogen, for which these transitions more closely correspond to the Franck–Condon principle.

Let us go back to the case of Ag_2 which has been documented well because
not only are ZEKE spectra available, but also the Rydberg series of Ag_2 have
been measured and assigned[209]. These two types of results provide an interesting comparison. In Fig. 20.8, the level diagrams of Ag_2 in the X state and
B state as well as the ionic state are shown. The equilibrium distances, ω_e and
the anharmonicity $\omega_e x_e$ have all been measured. A great deal of this information has been gleaned from Demtröder's group's analysis[209,210].

Now let us look at Ag_2 from the viewpoint of a more generic description
of channel interactions, rather than auto-ionization. The ladders, when
closely spaced, are expected to degrade towards the lowest values of $v^+=0$
for all excitations in v', which is a result of the overlap of the various vibrational progressions. We also expect the spectral profile and the individual
vibrational band intensities to be anomalous. The width of the individual
band also is expected to be anomalous. The experiments on Ag_2 with
ZEKE spectroscopy demonstrate that these bands are quite unusually

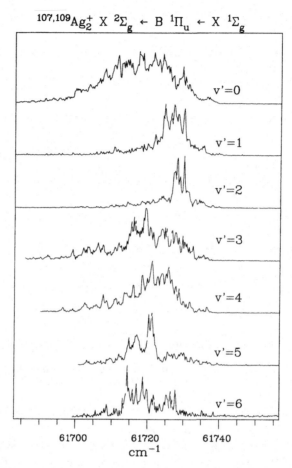

Fig. 20.9 The silver dimer. The spectral profile of each vibronic transition starting from different vibrations ν' in the B $^1\Pi_u$ state. The correct transition is at the blue edge of the band. The width is given by the rotational energy distribution.

broad (Fig. 20.9). This is not lack of instrumental resolution and is indeed much larger than the instrumental resolution. This broadening effect is real and is interpretable as precisely the multichannel coupling effect which is due to the rotational envelope of the originating states, as seen above. This is, of course, inhomogeneous broadening, which could be resolvable at higher resolution. Again, one must distinguish between the types of channel couplings which are involved.

These spectra can be understood in terms of these channel couplings which naturally lead to these types of inhomogeneous broadenings. The coupling which dominates the spectrum arises from the accessible coupled

$$X\,{}^2\Sigma_g^+ \; \leftarrow \; B\,{}^1\Pi_u\,(v'=4) \; \leftarrow \; X\,{}^1\Sigma_g^+\,(v''=0)$$

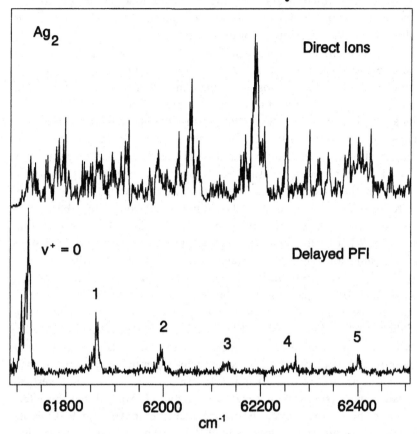

Fig. 20.10 The silver dimer. This again shows the direct ion current spectrum versus the ZEKE spectrum for the same region of the spectrum.

Rydberg series. These Rydberg states couple effectively because not only does each vibrational state have its own, private Rydberg series, but also each rotational state has its own Rydberg series. One again can compare this with the spectrum measured in ordinary Rydberg transitions by the direct ion current, whereby this system can be assigned to a whole series of various Rydberg ladders. The ionization threshold predicted by Demtröder's group[209] agrees perfectly with our ZEKE spectra, thereby confirming both measurements. Our vibrational ZEKE spectrum of Ag_2 is shown in Fig. 9.10. A detailed comparison with the direct ion current is given in Fig. 20.10. There are many interesting correspondences, even though the connection between direct ions and the ZEKE intensities is not easily visible. If you look at a particular band, it is much easier to recognize definite similarities between these spectra. Important differences due to the differences in the techniques, however, remain.

These differences are seen most directly by the detailed assignments of the direct ion spectrum in Fig. 20.11[209]. The assignment of Demtröder's group here shows the principal quantum numbers of the various series. For example, it is interesting to note that, on going from $v^+=4$ to $v^+=8$, one finds very strong transitions starting from $v'=4$ in the intermediate state. These are not strong in the ZEKE spectra, again indicating that channel coupling to the lower quantum numbers is effective, which is a peculiarity of ZEKE spectroscopy. In other words, one finds a very strong transition to $v^+=8$ in the direct spectrum. We do not find this strong transition in the ZEKE spectrum, as expected. It should not be as strong simply because of the long lifetimes of the competing ZEKE states at lower quantum numbers. What Rydberg spectroscopy has given us therefore is the assignment of auto-ionizing Rydberg states from $n=10$ to 25, extending to lower values of $N=4$–9. From this information, the auto-ionization structure that is operative for this particular ZEKE state can be derived. For the case of dense Rydberg manifolds (i.e. heavy molecules), essentially all the intensity of the pulsed field ionization (PFI) ZEKE signal comes from the indirect, below-threshold channel coupling crossing to manifolds with higher n at lower quantum numbers v^+. The system may prefer to have a ZEKE transition indirectly to the highest n and lowest v^+, this being a non-vertical transition populated via a lower n and a higher v^+. We feel that this effect is responsible for the apparently strange intensities in Ag_2 (Fig. 9.10). Note that, for all originating states in Fig. 9.10, $v^+=0$ has the highest intensity; it is clearly not a Franck–Condon process, but is understandable as the logical final state in a complete channel-coupling scheme.

The rotational level scheme for ZEKE states is displayed in Fig. 20.12. As a result of the temperature of the jet and the concomitant transitions to the B state of Ag_2, only the levels populated up to $J'=26$ are considered. This is indirectly a result of the large moment of inertia of Ag_2, the B value being $0.045\,cm^{-1}$. This means that levels up to $N^+=26$ in the X state of Ag_2^+ levels are populated according to the selection rules. Each of these rotational states again has its own Rydberg series energetically below that state. This is often referred to as Hund's case (d) or sometimes called the inverse Born–Oppenheimer approximation. The same rules as before apply, i.e., systems always mix and seek out states of higher n and lower quantum number N^+, and therefore transition occurs to the left only. These states of

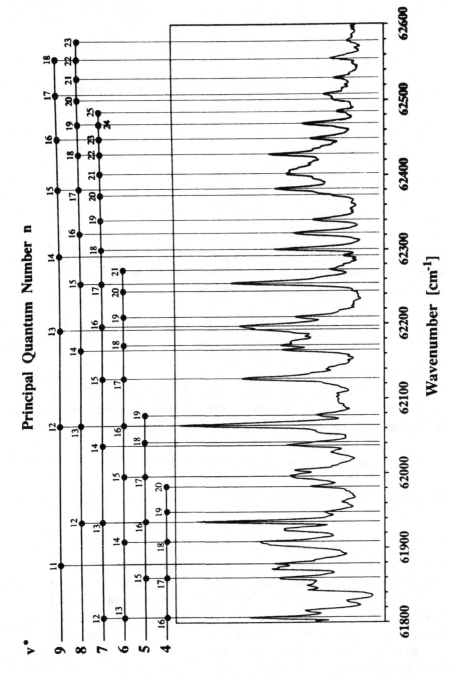

Fig. 20.11 The silver dimer. Direct absorption[209] showing high intensities for high vibrational states, in contrast to ZEKE spectroscopy.

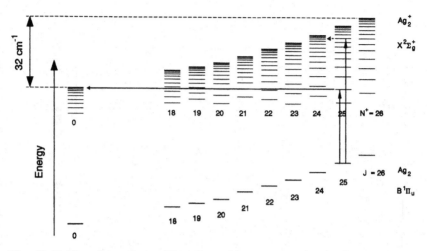

Fig. 20.12 The silver dimer. The absorption spectrum in the inverse Born–Oppenheimer representation. Here each rotation has its own Rydberg series. The rotational distribution populates up to $J=26$ in the B state. Correspondingly, in the strong channel coupling limit, all vertical transitions are red-degraded by this distribution.

high n can be coupled to lower N'^+ states under conservation of angular momentum

$$N^+ + \ell_n \to N'^+ + \ell'_{n'}$$

where ℓ_n is the orbital angular momentum of the originally excited Rydberg state with principal quantum number n. This coupling also must obey the law of conservation of energy, which, in the field-free case, is given by

$$R\left(\frac{1}{n^2} - \frac{1}{n'^2}\right) = B^+[N^+(N^+ + 1) - N'^+(N'^+ + 1)]$$

or for $\Delta N^+ = -1$

$$R\left(\frac{1}{n^2} - \frac{1}{n'^2}\right) = 2B^+ N^+$$

where R is the Rydberg constant and B^+ the rotational constant of the ion. In the field-free case only low-ℓ states are optically accessible. The range of rotational channel coupling here is restricted by the conservation conditions mentioned above since large negative changes in N^+ would eventually require large changes in ℓ_n, which is not possible by an intramolecular mechanism in the absence of the field. This is due to the lack of penetration of the core by the high-ℓ electrons.

In the presence of a field this restriction is removed since the linear Stark effect is now operative, which means that all these new Stark states have mixed ℓ character, hence permitting coupling of all ℓ quantum numbers. Of particular importance here are the low-ℓ states due to their penetrating character[112,120]. It must be remembered, however, that the field required is a sensitive function of n, the Rydberg state, the higher n the less field required for mixing.

Continuing this procedure to $J=1$, going to $N^+=1$ and taking all transitions together, we populate a higher band of width 32 cm^{-1}. If the lower band is also 32 cm^{-1}, all absorptions would be at the same energy and classically we should expect only a single line. By the principles mentioned above, however, the degradation to lower N^+ also assumes the eigenstate signature of the degraded state. This means that pumping $N^+=24$ at lower n populates increasingly lower N^+ at higher n. Hence, the absorption band has its blue edge at $n=150$ of $N^+=24$, but degrades to the red to $N^+=23$ at $n=250$, which leads to $N^+=22$ at $n=300$ and so lower N^+ and higher n are populated. As a result of this channel-coupling effect, all rotational transitions that should be at one energy are red-degraded, thereby showing peaks at lower energy just as if these states had been populated directly. This is seen in Fig. 20.13 with some intensity fluctuations that reflect changes in Rydberg densities for the relevant manifolds. The picture of channel coupling results from a level scheme such as that produced from the breakdown of the Born–Oppenheimer (BO) approximation, i.e. it is most readily seen in the representation of the inverse BO picture (Fig. 20.12).

In fact, we have here on this basis of extreme coupling a red-degradation of the rotational states with a band width approximately determined by thermal excitation in the ground state, a 32 cm^{-1} feature. This is exactly what is seen. The Rydberg states being pumped are converging into the $N^+=25$ level and from this the typical red-degradation to all quantum states at lower N^+ is observed. If you have the whole retinue from 1 to 26 for the J levels and the $N^+=1-26$, then you have to fold this into all the other Rydberg states of lower N^+. This is just another example of the degradation to lower quantum states typical for ZEKE spectra.

One of the useful aspects of the foregoing inverse BO level scheme (Fig. 20.12) is this prediction of the degradation of the spectrum by the width of the rotational energy distribution. This has the same origin as the $v^+=0$ preponderance in the vibrational spectrum. This means that the preferred intensity in the spectrum is not determined by the value of n under conservation in quantum number, but rather by the value of an optically pumped lower n that still couples to the highest possible n of a

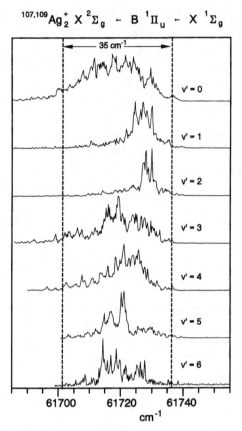

Fig. 20.13 The silver dimer. The rotational distribution curves some 35 cm^{-1} in the B state. This is then roughly the red-degradation in the ionic X state.

neighbouring series. It must be remembered again that ΔN selection rules are the same for the whole stack. Clearly the coupling of all n stacks is so strong that this rather than conservation in the quantum number dominates the picture. The latter would be the obvious choice in the Born–Oppenheimer frame. The frame used here in Fig. 20.12 is natural for the inverse Born–Oppenheimer (IBO) frame, which is more useful here, as evinced by the observed strong coupling of Rydberg stacks. In the normal BO picture we would also obtain a distribution of intensities as given by the thermal distribution, but the intensities would not be the strongest for the highest wavelength, this rather following a normal Franck–Condon picture. This is clearly violated for Ag$_2$, showing the usefulness of this different level scheme here, a picture that may well be appropriate for other ZEKE states.

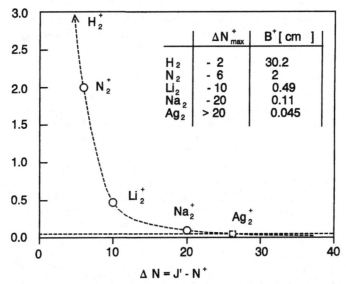

Fig. 20.14 Rotational auto-ionization for various diatomic systems as a result of forced rotational auto-ionization.

This type of behaviour is observed to a limited extent in other systems (Fig. 20.14) and depends, of course, on the rotational angular momentum involved. If you simply plot the width of this rotational mixing as has typically been observed for the series of systems hydrogen, nitrogen, lithium and silver, the bandwidth decreases with moment of inertia, which reflects the influence of level spacing on coupling.

This means that the coupling can be of some importance for ZEKE spectra. The band width is always expected to contain as its maximum the red-degradation due to the thermal excitation in the ground state which is due to the channel-coupling process. Hence, in utilizing spectra one wants to look not at the width of each peak, but rather for the origin which always appears at the blue edge for all peaks. It is the blue edge which indicates the transition of interest. One should therefore be able to see what the assignments are on the basis of our ZEKE results. For the case of this Ag_2 system, the results are presented in Table 20.1. If we take the centre of gravity of each band, the results are only adequate. If we compare the expected blue edge of our ZEKE results with the very-high-resolution work of Beutel *et al.*[209], we find that the origin at 61 747.1±3 cm^{-1} is observed. This needs to

Table 20.1. *Spectroscopic constants for the ground state of the silver dimer ion[209,210]; the spectroscopic constants shown are the ionization potentials (for the ionic ground state* $X\,{}^2\Sigma_g^+$ *of* ${}^{107,109}Ag_2^+$*) and the vibrational constants* ω_e *and* $\omega_e x_e$

v^+	IP(v^+) (cm^{-1}) blue edge[205]	IP(v^+) (cm^{-1}) Beutel *et al.*[209,210]
0	61 746.1±3	61 747.1±3
1	61 881.7±3	61 881.9±3
2	62 016.6±3	62 015.7±3
3	62 148.3±3	62 148.5±3
4	62 279.0±3	62 280.8±3
ω_e	136.2±0.4	135.8±0.3
$\omega_e x_e$	0.65±0.18	0.50±0.02

be compared with 61 746.1±3 cm^{-1} for the blue edge of the ZEKE spectrum, which is in excellent agreement and well within the experimental errors. This shows that, for this kind of heavy system, we need to take the channel couplings into account for fixing the spectroscopic origin of the transition. The centre-of-gravity results are outside the margin of experimental error. We also find that, if we stay with the blue edge, a better value of ω_e can be obtained. Even the anharmonicity is measured more accurately with the blue edge. I hope that this shows that the blue-edge-type auto-ionization is a reasonable explanation for complex systems of this type and affords an excellent method for assigning ZEKE spectra.

In summary, one can say that channel couplings, whenever they are possible, allow transitions to lower quantum numbers in ZEKE spectra and higher Rydberg states n, which consequently are optically allowed by the 'back door'. These are then lateral transitions to higher n Rydberg states. In general, when the couplings permit (i.e. the Rydberg level density is high enough), all systems in ZEKE spectroscopy seek their highest Rydberg states n. In general this can start as early in energy as is permitted by the thermal distribution function. One can also start at higher energy. At higher energies the Rydberg levels become denser, leading to improved coupling to the neighbouring Rydberg stack. Hence, coupling to highest n always takes place as soon as the spacing of the Rydberg levels permits; this need not reveal the IP, but rather a partially excited state of the ion. This means that the lowest higher vibrational or rotational quantum number in the ZEKE spectrum is observed in the extreme only. All these transitions appear at energies corresponding to the eigenstates associated with these lower quantum numbers. This process has the appearance of circumventing selection rules and violating the Franck–Condon principle. The transition to lower quantum numbers is associated with a higher value of the Rydberg state n and hence a longer ZEKE lifetime, which ensures its preponderance. For this to work, one must populate a Rydberg state that starts with a higher quantum number. A higher quantum number tends to be populated from below, whereby this red-degradation can proceed. To obtain high eigenstates at higher levels that are forbidden, one needs to start at high eigenstates at lower levels and to scan upwards. If one carefully selects the eigenstates at lower levels, one can use the Franck–Condon principle to assign the higher levels attained. In fact, it is feasible to walk around the Franck-Condon surface, thereby changing grossly the relative intensities of the bands. In general, as a matter of experience, one finds that one observes virtually all energy eigenstates at higher levels, be they forbidden or not, up to some 1000 cm^{-1} in the ion. This is reasonable in view of the mechanism described above.

This general view has to be extended to all ionic states that, of course, lie above the IP. Far more importantly, however, these ZEKE states are all very long-lived even though they are embedded within the ionization continuum. It is this new effect which makes ZEKE work so well at high energies in the cation system. It is these kinds of ZEKE states and the extreme duration of their lifetimes (in the 20–100 μs range) which was at first quite astonishing and in conflict with common wisdom. It was generally thought that such states, having energies well above the IP, are all very quickly auto-ionized into the continuum and eject an electron on the femtosecond timescale. One knows now that, even though these states have plenty of energy to ionize, they do not simply do so because this electron is sitting in a very-high-ℓ and m_ℓ orbit, orbiting around the core, but staying away from the core. This leads to a spectator-type state. One can consider the high Rydberg states embedded within the continuum to be converted quickly into high-ℓ and m_ℓ ZEKE states, thereby making them into spectator states such that the electron is largely uncoupled from the core. One can consider these as special islands of stability within the ionization continuum.

In summary, it may be useful to consider a 'front door' and a 'back door' mechanism in ZEKE spectroscopy. The former is analogous to optical spectra that are then transformed into long-lived ZEKE states. Here typical Franck–Condon factors apply and *ab initio* calculations work best. Fields are now only detrimental (cf. Chapter 21). The 'back door' mechanism is a new path not available to conventional spectroscopy. It allows many new states to come into view that are not seen with standard, i.e. Franck–Condon-limited, techniques. It is the place where *ab initio* calculations will be less successful, but MQDT might be helpful[211]. It is the place where small fields help (cf. point 6 above), although larger fields still lead to a loss of signal. If one of the channels is dissociative or leads to ionization due to its being at lower n, it will even open up a path to ionization, since, in spite of the system normally going to high n, an exit channel leads to an increased phase space and dissociation or ionization[212].

21

Mixing of states by fields

The foregoing discussion clearly shows that channel coupling between Rydberg series is one of the important new mechanisms in ZEKE spectroscopy, leading to many new transitions. Also, from the rotational levels it is clear that the coupling reaches more levels through more stacks of Rydberg series as the molecule becomes heavier and the rotational constants decrease. Hence level spacing is an important parameter in deciding how many Rydberg series are coupled. This is also demonstrated by the fact that, for light systems, ZEKE spectra appear to correspond more nearly to Franck–Condon rules, whereas heavy systems need not. This is due to the fact that heavy systems, owing to their increased number of levels, are increasingly capable of drawing intensity from the neighbouring series. It should be remembered that the excitation process, in general, will be considered to follow Franck–Condon rules, but for heavy systems this process might involve higher vibrational ground states, leading to a Rydberg ladder associated with a higher vibrational state of the ion. Thus channel coupling to a lower vibrational state of the ion at a concomitantly higher Rydberg state n is favoured. This then appears to be a violation of the Franck–Condon rules. Two-photon excitation is here often of advantage since high state selection in the intermediate level avoids ambiguities about the method of excitation. The same is true for complex molecular systems that do not have such strict selection rules and hence will prefer to undergo the Franck–Condon transitions.

One can ask how far such coupling can proceed. Clearly this must also depend on the number of states of the various series, as well as their mixing. This, in turn, depends on the field. In a hydrogenic system one has well-defined levels described by the Rydberg formula in terms of the binding energy of the Rydberg electron as

179

$$E_H = \frac{-R_H}{n^{*2}} \quad (R_H = 109\,677.58 \text{ cm}^{-1})$$

For non-hydrogenic systems, particularly for molecules, this formula must be modified so that $n^* = n - \mu'_\ell$, which includes the quantum defect μ'_ℓ. This quantum defect will be different for different Rydberg series. In general, it will be largest for s and p series, less for d and f series and negligible for high ℓ. In a hydrogenic system any electric field suffices to induce a linear Stark effect (i.e. full ℓ-mixing and therefore approximately n-fold lifetime enhancement), no matter how small this field is.

One may thus consider that the higher ℓ states are hydrogenic in character, but that the first few ℓ states are lower in energy due to the quantum defect. As the field is increased one goes into parabolic Stark states as eigenstates of the system having mixed ℓ character. At a critical field all these ℓ states including the few low-ℓ states are suddenly mixed and there is a sudden jump in the lifetime. For a given field strength there will be a value of n at which all levels couple to form broadened states. The general field at which this broadening is expected in a hydrogenic picture is at

$$F = 1.71 \times 10^9 n^{-5} \text{ V cm}^{-1}$$

which is the Inglis–Teller limit[213,214]. This is a point at which the broadening of states equals the level spacing above. At this point all states are smeared out and all transitions are possible. At this point mixing produces migration to high-ℓ states and this becomes much more facile, hence enhancing the mixing of ℓ states. This is clearly seen in the work of Vrakking and Lee for nitric oxide[128,215,216] in Fig. 21.1, in which one sees a clear break in the lifetimes with energy, indicating that, above a certain n, this level smearing opens up the channel crossing to long-lived ZEKE states.

In the presence of a field this onset is seen to change. It must be cautioned that this formula is hydrogenic in origin and so the real situation could be different for molecular systems[37,215,217]. It must be mentioned, of course, that this onset is not unique but rather depends on the accidental locations of the zeroth-order levels that are coupled between the stacks, which makes them energetically quasi-degenerate. These accidental degeneracies were observed by Vrakking and Lee[128]. It must be emphasized that the above discussion, even though it applies to xenon, relies on the hydrogenic picture, which is also the basis for the simple Inglis–Teller formula stated above.

The case is more complicated for a non-hydrogenic system. The field-free Rydberg states for this non-hydrogenic case have well-defined ℓ character with lowest-ℓ states split off in energy from the high-ℓ manifold. As a result,

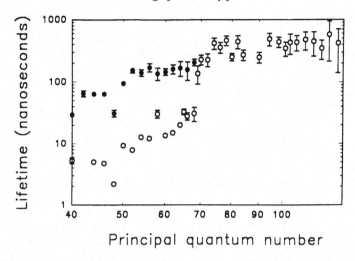

Fig. 21.1 Nitric oxide. Lifetimes of Rydberg series from the $(A'\Sigma^+ N_A=1, J_A=\frac{1}{2})$ intermediate state, determined using delayed pulsed-field ionization. The figure shows the lifetimes for an f series converging on $N^+=1$. Open circles represent measurements under the minimum DC electric field condition (25 mV cm^{-1}), whereas filled circles represent measurements with appropriate DC electric fields applied to observe an enhancement of the lifetime. This demonstrates the jump in lifetime by factor of ten at $n=70$ at low fields. At higher fields this increase in lifetime disappears[215].

there exists a critical (a sub-Inglis–Teller) field for the onset of the linear Stark effect and consequent lifetime enhancement in a non-hydrogenic system. At a field below this critical value, there is no significant ℓ-mixing, so that the optical excitation populates a short-lived well-defined low-ℓ state. In the presence of a field above the Inglis–Teller limit, the Rydberg states are the totally ℓ-mixed hydrogenic-like Stark states (hence, they are longer lived). This critical field depends on n and on the quantum defect μ'_ℓ in a given ℓ state. The mixing starts when the field shift in the Stark states at high ℓ is larger than the zero-field splitting. For $\ell=0, 1$ ($n=2$) this is the simple case of sp^1 hybridization in chemistry. This, of course, again decreases as the quantum defect decreases with increasing ℓ roughly as ℓ^3.

When a particular Rydberg state n is realized, this can lead to values of ℓ up to $n-1$, although low-ℓ states are initially populated by the laser. Correspondingly, m_ℓ states are possible and could exist in the range $m_\ell = -\ell, ..., 0, ..., +\ell$ leading to a $2m_\ell + 1$ degeneracy. Now, in an external field, one has new Stark states and ℓ is no longer a good quantum number in the cylindrical external field. This will produce mixing of the ℓ states. Such mixing is already substantial at stray fields of 10–20 mV cm^{-1}. If, in addition, this field is not rendered cylindrically symmetrical by virtue of a neighbouring charge[119] there will be a corresponding loss of m_ℓ as a good quantum number. The presence of any ions now proceeds to mix m_ℓ as well as ℓ states and facilitates the coupling to form ZEKE states with high ℓ and m_ℓ. This will then modify the intensity and lifetime in the ZEKE spectrum since this coupling process is important in the formation of ZEKE states. Whatever the detailed nature of this coupling between low-ℓ, m_ℓ and high-ℓ, m_ℓ states, one must remember that these processes are subject to microscopic reversibility, i.e.

$$\text{mixed } \ell \text{ states} \rightleftharpoons \text{mixed } \ell, m_\ell \text{ ZEKE states}$$

Hence, any mechanism that favours production of high-m_ℓ, ℓ states also favours their loss until an equilibrium has been attained for small fields and ion concentrations. However, it must also be remembered that, owing to the mixing of the m_ℓ components, one expects close to a $(2m_\ell + 1)$ degeneracy. Hence the phase space for high-ℓ states is much increased. This explains the stable formation of high-m_ℓ, ℓ ZEKE states in the fields in spite of microscopic reversibility. Such field effects are typically present in all but the most thoroughly shielded experiments. A typical stray field[214] will be of the order of 20 mV cm^{-1}. Extreme methods can reduce these fields to some 100 μV cm^{-1}. This may, however, not be desirable since ℓ mixing and hence the ZEKE effect depend on such interactions. Experiments on xenon atoms[216] have demonstrated that added ions do indeed modify the ZEKE intensities in transitions to the $^2P_{1/2}$ state above the ionization energy. Experiments on molecules such as benzene, however (Fig. 21.2) at first did not display the effect that added ions do change a ZEKE spectrum. The spectrum (c) in the presence of added ions from (b) is identical to the pure spectrum (a); the difference is shown in (d). This does not mean that the field effect is not operative in molecules, rather it implies that, under the conditions of a typical experiment, we do not require the added ions for this mechanism to operate[218]. More recent experiments[219] have, however, shown that, at very much lower ion concentrations in benzene, the ZEKE signal also decreases. In fact, extrapolating these data appears to indicate the necessity of ions in order to produce ZEKE states from ℓ-mixed Rydberg states. Under some conditions the latter are produced initially by the stray fields. This reconciles the experiments of Alt *et al.*[130] and Vrakking *et al.*[220] in that, for typical molecular experiments, we have enough ions for there to be an equilibrium between Rydberg and ZEKE states.

Hence very small stray fields are clearly seen to enhance coupling between Rydberg series for atoms or molecules. These fields are essential in

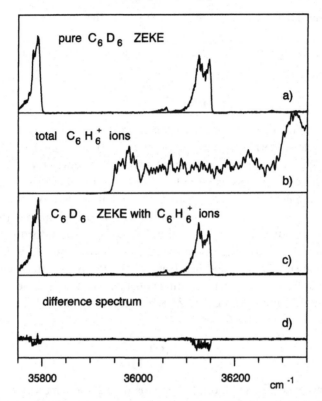

Fig. 21.2 The influence of ion density on benzene ZEKE spectra. This shows that ZEKE intensities are unaffected by the presence of extra ions due to the other isotopomer.

some form to produce the requisite ℓ mixing, m_ℓ mixing to produce ZEKE states being produced by ions. This coupling is here described as the effect producing the mixing. As more members of the Rydberg series are mixed this produces inhomogeneous broadening, i.e. maintaining in principle sharp transitions, particularly if high-n and low-n transitions are coupled.

Hence we can obtain an overall mechanism for the production of ZEKE states that proceeds as follows. In an initial step of photoexcitation we form Rydberg states of low ℓ:

$$M \overset{h\nu}{\rightarrow} M \text{ (R) low } \ell \tag{1}$$

In the presence of even a small residual field of some $10–20\,\mathrm{mV\,cm^{-1}}$, high-$n$ Rydberg states are mixed in the corresponding Stark fields so that all states are strongly mixed. This is a property of high-n states that ensures this mixing. Hence we have produced Stark Rydberg states with strong ℓ mixing.

In the presence of a DC field during irradiation, these Rydberg states decay:

$$M(R) \overset{\alpha(F)}{\to} \text{decay} \tag{2}$$

by way of auto-ionization or pre-dissociation. At low fields the function is linear in field. Alternatively, in the presence of other ions these Rydberg states undergo m_ℓ mixing. For this the cylindrical symmetry of the field must be broken, a process that is only possible via the inhomogeneous component of the field exerted by the ions. This will now tend to stabilize high ℓ, m_ℓ values of the electron. The electron now orbits around the core keeping its distance from the core, thus becoming a spectator to the core and considerably reducing its interaction with the core. This leads to considerable stabilization and longevity of the Rydberg state – which we now term a ZEKE state. This permits considerable delays in field ionization before splitting these neutral ZEKE states into ions and electrons for detection:

$$M(R) \overset{k_1[M^+]}{\to} M(Z) \tag{3}$$

In the absence of a field, or at low fields with some ions, step (2) is not operative relative to step (3) and all Rydberg states within the 8 cm^{-1} ZEKE bandwidth of n that can couple produce ZEKE states. Hence we have 'saturation' and no further increase in [M$^+$] enhances the yield, as observed by Alt. Conversely, in the presence of a field, ion-dependent intensities will be observed. If the ion concentration [M$^+$] is increased even further we observe a widening of the ZEKE bandwidth from 8 cm^{-1} to some 12 cm^{-1}, hence a loss of resolution. This means that the ZEKE bandwidth includes states with lower Rydberg quantum number n. These Rydberg states are always populated optically, but normally do not convert to long-lived ZEKE states and hence are lost to detection. Since, as in (3), ions are effective in producing these ZEKE states, we conclude that, with sufficient concentrations of [M$^+$], the inhomogeneous field is sufficient even to convert some lower n states into ZEKE states, thus widening the ZEKE band. At higher fields, (2) becomes operative rather than (3), destroying the signal. This can be compensated by an added ion concentration [M$^+$]. Similarly, [M$^+$] operates only over a small bandwidth since, analogously to the stepwise change in lifetime with fields, we here have a sudden change with ions.

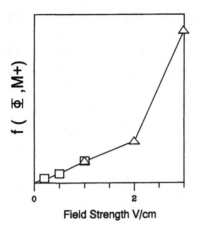

Fig. 21.3 The field dependence of the quantum yield of ZEKE formation. The plot shows the left-hand side of equation (5) versus the electric field strength.

Define a quantum yield for ZEKE formation $\phi(Z)$ as the ratio of the in-field to out-of-field signals

$$\phi(Z) = \frac{I_{\text{in-field}}}{I_{\text{out-of-field}}} = \frac{k_1[M^+]}{k_1[M^+] + \alpha(F)} \tag{4}$$

which follows from (1)–(3) above. Hence

$$\frac{1}{\phi(Z)} - 1 = \frac{\alpha(F)}{k_1[M^+]} \tag{5}$$

Now (5) produces a unique prediction of the quantum yield for ZEKE formation for defined reduced variables $F/[M^+]$ (Fig. 21.3). To demonstrate the competition between the two effects above for the case of benzene we expect equivalence when $\alpha(F) = k_1[M^+]$ and hence to have an equivalence point of $\phi(Z) = 0.5$. This occurs with a concentration of only six ions per mm^3 at a $1\,V\,cm^{-1}$ field strength. Hence about 100 ions per mm^3 would lead to a 'saturated' signal. Typically we find no field effects below $0.2\,V\,cm^{-1}$, i.e. the same signal for DC or delayed fields. Effectively this means that all these experiments contained at least 20 ions per mm^3.

It is interesting to note that similar effects have recently been seen for Ar atoms. Here Hsu et al.[39] saw no $^2P_{1/2}$ signal in synchrotron experiments in the presence of some field. Hepburn, however, showed[221] that this signal could be restored by the addition of ions. Indeed, the signal is present in the absence of a DC field. This is in accord with the above mechanism and constitutes a general note of caution for synchrotron experiments in particular. It means that very careful shielding of the photo-ionization region is required for synchrotron experiments.

It is not yet entirely clear whether mechanisms not based on ions cannot also form high-m_ℓ ZEKE states. Clearly microwave fields are effective here[222].

22

The effect of a field on the ionization potential

The theoretical effect of fields in ZEKE spectroscopy was initially analysed in the classic work of Chupka[62] and Bordas and Helm[113]. The extraction of the ZEKE states is done with a pulsed delayed field, although delaying would suffice even without pulsing. One should be aware that, if one imposes a field, one lowers the ionization potential (Fig. 6.5) due to the electric field. Hence any experimental IP value is field-dependent. A text-book example of this classical effect is the elementary field-ionization effect. This can be used to measure ionization potentials by plotting the onset of ionization as a function of $\sqrt{F}$, where F is the field in V cm^{-1}, and extrapolating to $F=0$ (Fig. 22.1). Employing typical values of $F=100$ V cm^{-1} produces a straight line of gradient 6 $\sqrt{F}$. We have done this for a couple of cases and here show that, with benzene, we can demonstrate that such an extrapolation is possible (Fig. 22.1(a)). It gives the correct IP. This is the case of a large constant field being present at ionization. This process is also referred to as adiabatic ionization.

One has to remember that all these are transitions within the Stark manifold produced by the presence, intentional or stray, of an electric field F. This Stark manifold is present for each Rydberg level n and broadens this level. Ionization is then from these broadened, Stark-split states. This has been reviewed well for atoms by Gallagher[111].

These Stark-split states cross and form avoided crossings in non-hydrogenic systems. Typically, a rapid rise of the electric field results in preservation of the Stark state character. In particular, the dipolar moment of the state along the field direction is nearly constant during the entire passage from a low field to an ionizing field. This kind of passage to ionization is hydrogenic-like and is known to have the ionization onset at a binding energy of E (cm^{-1}) $\approx 4\sqrt{F}$ (V cm^{-1}). The field-shifted energies of Stark

187

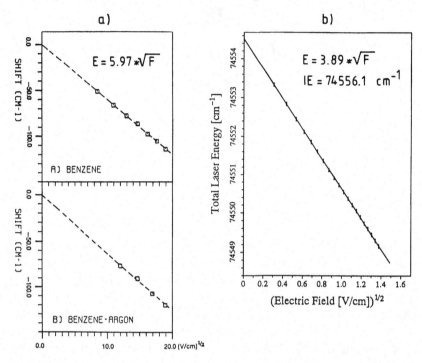

Fig. 22.1 Experimental observation of (a) the lowering of the ionization potential with field, starting with low-n states and (b) the lowering of the IP measured with ZEKE spectroscopy at high n.

states go approximately linearly with the field and the energy curves for different Stark states cross.

Actually, in a non-hydrogenic system, a sufficiently slow rise of the field will result in so-called adiabatic passage to ionization. In this case each passage through an avoided crossing drastically changes the character of the state which is due to the migration jumping through the various diabatic states (Fig. 22.2). Fast and slow in the above discussion relates to the spacing and hence the splitting of the levels relative to the inverse of the time for the passage. The magnitude of the splitting in an avoided crossing decreases strongly with n, the Rydberg state. Hence, for a given rate of slewing of the field, high-n states typically manifest diabatic passage to ionization whereas low-n states typically manifest adiabatic passage. This is manifested in $4\sqrt{F}$ versus $6\sqrt{F}$ behaviours, respectively. For this reason ZEKE states that exist only at high n typically manifest diabatic passage to ionization.

One should remember that both adiabatic and diabatic passages can

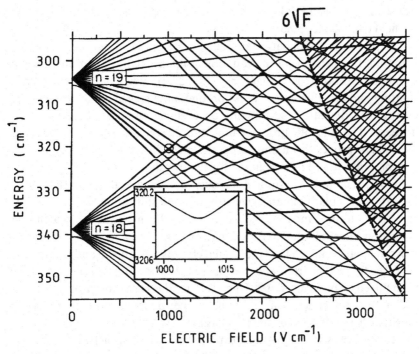

Fig. 22.2 The adiabatic route to the ionization threshold[223].

occur at the same time[224], owing to the accidental near degeneracies discussed above. The $6\sqrt{F}$ results are generally the high-field excitation results (Fig. 22.1) from low n. Whatever the case, one can always extrapolate with $\sqrt{F}$ and obtain the correct ionization potentials, as we showed for the cases of benzene and benzene–argon complexes some years ago[225]. For ZEKE experiments this can be done with great accuracy. Most ZEKE experiments to date have been within the diabatic range, demonstrating a field dependence that actually is $3.85\sqrt{F}$.

When one looks at this problem in more detail one must first ask the question of the nature of the states excited. The Inglis–Teller limit for the atomic system's state suggests that, for a field of some 30 mV cm^{-1}, states above $n=140$ are mixed and states at lower n are unmixed. This implies a sharp onset for the hydrogenic system, which is, of course, an oversimplification. Most ZEKE experiments are carried out at $n=200$–300, at which 2 mV cm^{-1} completely mixes all levels. Since most experiments have stray fields of this order, one can presume that all ZEKE experiments start from mixed and Stark-split states. Thus pulsed-field excitation shows the field dependence $3.85\sqrt{F}$.

The problem of adiabatic and diabatic passages can be seen in a Stark manifold. In a general way, if you look at the usual Stark manifold from a standard textbook (e.g. Gallagher[111] (Fig. 22.2)), you have the usual Stark splitting and the adiabatic process produces the jagged curve crossing the curves. The diabatic case remains on a single curve as it decreases or increases. The adiabatic process from unmixed states gives a $6\sqrt{F}$ curve and the diabatic process from unmixed states gives the $4\sqrt{F}$ curve. A change of $1\,\text{V}\,\text{cm}^{-1}$ in some 10 ns is required for the transition.

In brief therefore, the static draw-out field has two effects on IPs. It distorts the molecular field and gives rise to a red shift of the IP: for $E=1\,\text{V}\,\text{cm}^{-1}$, $\Delta\text{IP}=6\,\text{cm}^{-1}$ and for $E=4\,\text{V}\,\text{cm}^{-1}$, $\Delta\text{IP}=12\,\text{cm}^{-1}$; and it broadens the ionization threshold due to Stark splitting. One has a lowering of ionization potential because a field is on. That effect is well known and the process is proceeding either from above or from below the Inglis–Teller limit, i.e. from mixed or unmixed states. One also broadens the ionization threshold and shifts the energies. In fact one has to go to new quantum numbers in the Stark field. Of course, one also has Stark splitting.

Low-field pulsed-delay ZEKE experiments are the practical method of choice, even for mixed states. This presents a shift in the onset but this is readily calibrated. Also a delayed field leads to sharper and, for that matter, more intense spectra. This intensity effect is quite important though not yet fully understood. In any case, one extrapolates with various delayed fields to zero to obtain an accurate energy for the state based on the photon calibration described above. This gives highly accurate ionization potentials, as noted above. One must realize that this is an extrapolation of field for mixed levels, hence with a slope of $4\sqrt{F}$. In summary, for a given stray field low-n states go as $6\sqrt{F}$ and high-n states go as $4\sqrt{F}$. The latter is the general rule for almost all ZEKE experiments.

One further aspect of interest for the distinction between diabatic and adiabatic passage is not just that the former is typical for the high-n ZEKE states and the later for low-n states, but also that the Stark states split into so-called 'red' and 'blue' states. In Fig. 22.2 the 'red' states go to lower energy with the field and the 'blue' states go to higher energies. These red and blue states differ in their stability against ionization. In particular, the threshold field $F_{\text{thr}}(n_1, n_2, m)$ (which is a function of the parabolic quantum numbers n_1, n_2 and m and was computed by Baranov *et al.*[226]), is 2–3 times larger for the bluest Stark states than it is for the reddest state. In principle, the decay of the Stark states in the presence of an electric field is possible by the tunnelling through the saddle point of the combined Coulomb plus DC electric field potential. However, at the high n values of interest in ZEKE spectroscopy, the tunnelling rate of the states under below-threshold conditions ($F<F_{thr}(n_1, n_2, m)$) decreases so fast that the tunnelling mechanism can be neglected for all practical purposes[226]. Hence, the Stark state either ionizes after some few nanoseconds if it is above the so-called hydrogenic threshold ($F>F_{thr}(n_1, n_2, m)$) or it remains bound otherwise. In the presence of an applied field F, all the highest-n manifolds with the field-free binding energies Ry/n^2 (cm^{-1}) $\leqslant 3\sqrt{F}$ (the field F is in V cm^{-1}) undergo complete ionization (both red and blue Stark states are ionized). However, in the energy range between about $3\sqrt{F}$ and about $4\sqrt{F}$ there is only a fractional ionization[227] such that mostly the labile red states ionize, whereas the more resilient blue states survive the electric pulse. The surviving states can be selectively detected by ionization in the presence of a second field F_2 of opposite polarity and the same or even smaller magnitude, $|F_2| \leqslant |F_1|$, than the first field F_1[227]. Now the blue states become red states, which ionize readily. The simple physical argument for this is seen by reference to Fig. 6.5. The blue states are harder to ionize because the electron sits on the left-hand side, away from the saddle point. The red states, on the other hand, are adjacent to the saddle point[135] and hence ionize easily. Hence also, a reversal in polarity reverses the states. By looking at either only red states or only blue states, one can sharpen the ZEKE transitions.

It should be noted, though, that blue states can also be converted to red states slowly by themselves, simply by the electron being scattered from the core, hence converting it to a red state. Hence repetitive pulsing often produces an apparent small revival of the signal, even with the same draw-out pulse.

23

Intensity effects due to electric fields

It is observed that very small stray fields are instrumental in the production of mixed Rydberg ℓ states[37]. For typical experiments very small fields at typical ion concentrations below 200 mV cm^{-1} do not affect the ZEKE signals even if present at the start. On the other hand, somewhat larger fields of 3–10 V cm^{-1} lead to a decrease in ZEKE intensities (Fig. 23.1)[216], particularly if the field is present at the same time as the laser, namely under so-called in-field conditions. It at first appears puzzling how these ZEKE states are influenced. The answer is to be found in the optically pumped high-n, low-ℓ state, prior to its becoming a ZEKE state. Note that there is a rapid component of decay up to some 10 ns and then a slow component of 20–50 μs. This is shown in Fig. 23.2. After some 2 V cm^{-1} the intensity drops again, caused either by a shortening of the fast lifetime or by a change in population reducing the lifetime. This is reasonable in view of the fact that low-ℓ states have electron orbits that pass near the core, the distance going as $\frac{1}{2}\ell(\ell+1)$[116]. This makes these low-ℓ states particularly sensitive to applied fields, leading to their decay. This will produce fewer long-lived ZEKE states and hence a drop in signal as shown . This effect was first reported by Pratt[228] and has generally been reported in most ZEKE experiments since then. An alternative possibility is that fields change the long ZEKE lifetimes directly, but recent experiments by Held et al.[229] demonstrated that this simple explanation is not the case. These lifetimes are the same whether excitation is in the presence or in the absence of an electric field, even though the intensities are lower for the former case. This was also seen for the silver dimer (Fig. 23.3). This again is reasonable because high-ℓ states do not go near the core and hence are less susceptible to moderate fields.

A very low field together with ions induces the transition to ZEKE states and hence lengthens lifetimes. When the field increases above some 2 V

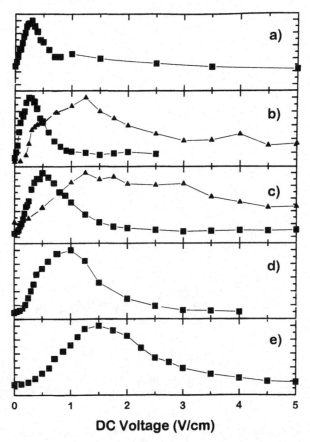

Fig. 23.1 Xenon. This demonstrates low-voltage enhancement of the signal and large voltage attrition. The dependence of the Xe non-pulsed field ionization signal, obtained at a fixed time delay of 80 ns between the laser excitation and the pulsed field ionization and extraction, on the ambient DC electric field is shown. Curves were obtained for excitation to (a) $n^* = 80$, (b) $n^* = 70$, (c) $n^* = 60$, (d) $n^* = 52$ and (e) $n^* = 43$. Squares represent excitation to an ns' level, whereas triangles represent excitation to an nd' level[216].

cm^{-1} it acts to reduce the lifetimes or intensities of these short-lived states, hence reducing both the number of subsequent states and the intensity of the final long-lived ZEKE states. This is reasonable since, for low-ℓ, field-induced interaction of the Rydberg electron in the neighbourhood of the core is much more prominent. The intensity of the long-lived ZEKE states at high ℓ is not expected to be influenced by the field and hence the long lifetimes of the experiment are the same with or without the field. The intensity change then is observed for the long-lived ZEKE states, but the

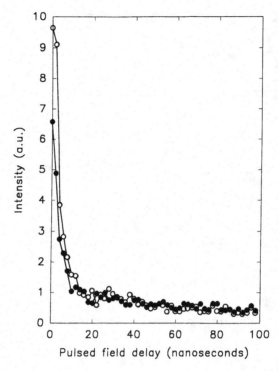

Fig. 23.2 Nitric oxide. The change in intensity with pulsed-field delay. Note the strong change in intensity in the first 10 ns demonstrating multiple decay times[215].

effect is exclusively due to lifetime or intensity changes in the short-lived optical Rydberg states and changes in optical population of the Rydberg state. This then leads to the experimental prescription of avoiding fields until after the light source has been turned off, for optimal ZEKE intensities.

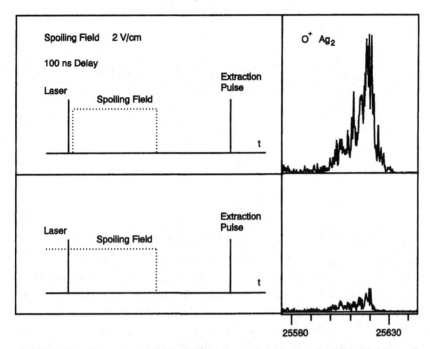

Fig. 23.3 The dependence of the ZEKE signal intensity on the initial field conditions. This demonstrates the typical decrease in intensity for in-field excitation. This effect is greatly reduced for fields of around $50 \, \mathrm{mV \, cm^{-1}}$.

24

Examples of systems studied

24.1 p-Difluorobenzene

As a further example of this spectroscopic method, consider a normal molecule like p-difluorobenzene and the best photoelectron spectrum reported to date utilizing small excess energy[31]. A very good photoelectron spectrum with about 5 meV resolution was achieved (top of Fig. 9.1) owing to the fact that a photon energy with only a very small excess energy was used. In comparison, the ZEKE detector measures something different (Fig. 9.1). This is shown in the ZEKE spectrum below the PES spectrum. The apparent difference is not just the resolution, it is also a question of there being a new detection scheme. There are several variants of this technique, as I have pointed out: just above threshold or just below threshold. All these ZEKE experiments are performed with two-colour excitation, but let me point out some features here. There are also certain problems with two-colour experiments. The beauty of a two-colour experiment is that you select a definite intermediate state on the way to the ZEKE state so you know where you start on the way to the ZEKE state (Fig. 24.1). This knowledge is essential for most ZEKE experiments. The bad news is that you always have interferences from multiphoton absorption of a single laser into the continuum. This effect creates background signals, mostly due to ions or accidentally excited states. To get the true ZEKE spectrum, you must always measure a difference spectrum with one of the lasers on or off and preferably delayed. This is a small but important experimental point. Of course, these are REMPI experiments in which a real state is located energetically between the ground and excited states. Absorbing the photon causes difficulties when the intensity needed goes through a virtual state and leads to multiphoton absorption in the ion and, hence, uncontrollable further excitation. Nitric oxide appears to be an exception to this general

196

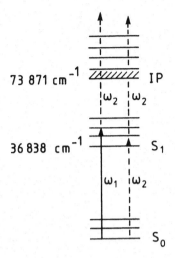

Fig. 24.1 One-colour versus two-colour two-photon excitation of *p*-difluorobenzene. Even in a two-colour experiment there is some inadvertent one-colour two-photon absorption. By slightly delaying the application of the second colour the error can be subtracted out.

and serious problem[230]. Here two-photon, one-colour excitation via a virtual level appears to work.

Let me illustrate what happens when one selects an individual intermediate state as the originating level (e.g. the vibrationless ground state of the S_1 of *p*-difluorobenzene) for the ZEKE experiment. This is the origin of our spectra in Fig. 24.2. The ZEKE spectrum displays very clear progressions, the 0–0 transition being very strong, since a ν_6 progression, a ν_5 progression etc. are seen, so you really can assign this spectrum very nicely. The Franck–Condon effect is used to our advantage since this provides the propensity rules which are not entirely obscured by channel couplings. In contrast, without such propensity rules any state that is energetically possible will be seen. Propensity rules are very useful for assignments. This is shown in another example (Fig. 24.3). Suppose that the first excited ν_6 state is the originating intermediate state. You notice that the transition to 0^0 in the ion becomes very weak and the progression in ν_6 becomes much stronger. This indicates, in other words, that one has many kinds of ZEKE states, but that the ν_6 mode in S_1 will emphasize the ν_6 mode in the ion. This is in contrast to the origin which does show a strong transition to 0^0 (Fig. 24.2). This becomes a very useful way to see the ν_6 progressions in the ion and hence to assign the cation state if the assignment in the initial S_1 state is known. So this becomes a very useful tool. Not all of the transitions are equal, as

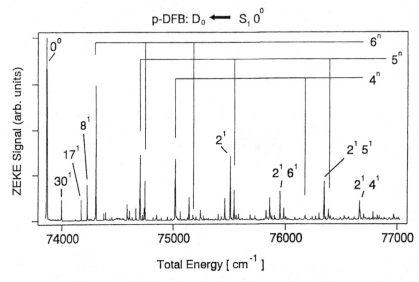

Fig. 24.2 Excitation of *p*-difluorobenzene with a second colour from the S_1 ground vibronic state to the lowest state of the ion. Note the strong 0–0 transition and the other vibronic propensities.

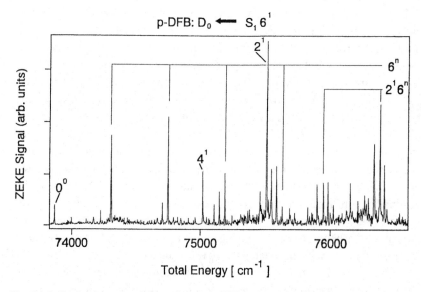

Fig. 24.3 Excitation of *p*-difluorobenzene with a second colour from the ν_6 mode of the S_1 electronic state. Note the decreased population of the ground vibrational state of the ion and the changed propensities.

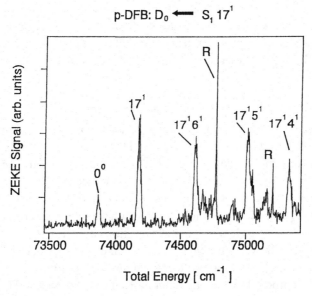

Fig. 24.4 Excitation of *p*-difluorobenzene with a second colour from the ν_{17} of the S$_1$ state. Note the change in propensities. The prominence of the ν_{17}^+ in the ion facilities vibrational assignments.

would be the case were channel coupling involved. The propensity rules that exist are very useful both for assignment and for bringing out weak transitions (Fig. 24.4). Perhaps the best example is shown in Fig. 24.4 in which ν_{17} vibration in the S$_1$ state of *p*-difluorobenzene is pumped. Notice a ν_{17} peak appearing nicely in the ZEKE spectrum. This becomes a useful method even to detect states that are otherwise hardly visible. The lesson then is that propensity helps in the assignment of the ionic state from the known S$_1$ assignments. Other transitions, however, are seen; even forbidden transitions. Essentially no transition appears to be invisible.

24.2 Cluster vibrations

Let me go on then to the spectra of clusters. In this particular case, we are going to have examples of clusters of phenol with water, with methanol and the like. At the start, let me point out that, if you have two different molecular systems that form a cluster, a blue one and a red one, you clearly have initially $3N_B - 6$ blue intramolecular states and $3N_R - 6$ red intramolecular states. When you marry these two together, you have, of course, $3(N_B + N_R) - 6$ states and, if you subtract this from the above, six intermolecular modes remain that were not present in either molecule before. These are the new states for which we are looking and which have been difficult to observe. Very few, if any, examples have been reported of complex systems in which all six of these modes can be seen or properly identified.

24.3 Phenol and its clusters

Before we get into phenol clusters, however, let us consider the simple situation of phenol as a monomer (Fig. 24.5). This spectrum already shows the OH vibrations nicely. You notice that the bending vibration δ_{OH} shifts significantly to lower energy in the deuterated molecule. The τ_{OH} vibration shifts somewhat, but not nearly as much. It also flips in energy to the other side of the ν_6 vibration. Secondly, what you have now is a series of vibrations, ν_{11}, ν_{16a}, ν_{18b}, ν_{16b}, which are a'' vibrations and should be forbidden, but are slightly allowed (enough for one to make an assignment). This is no doubt again that these data result from channel couplings.

Now let us proceed to add one water molecule to the phenol. The normal coordinate excursions can be seen in the usual arrow diagram showing the six possible intermolecular van der Waals modes (Fig. 9.6). Two different symmetries are involved. The spectrum contains all the various kinds of vibrations which can be assigned directly to van der Waals modes. What is interesting is that we also see the definite, but very slight, changes resulting from these van der Waals-type vibrations in the modes of the parent molecules. In Fig. 24.6, we give a pictorial (but not especially accurate) view of the clusters of phenol with various molecules. We can now proceed to pump ZEKE transitions via these new intermediate states as shown in Fig. 24.7 for the phenol–methanol cluster. The spectra show that we can actually excite into a whole series of van der Waals overtones and combinations, starting energetically at various initial van der Waals states as the intermediate state. At the ionization potential based on the 0–0 transition of the vibrationless level of the intermediate, we see the overtones starting from this origin. Exciting the η_1 in the intermediate state, this single van der Waals mode totally changes the spectrum, leading to a completely different Franck–Condon envelope. This is very helpful from the point of assignments. You can also walk around the Franck–Condon surface in ZEKE spectroscopy. In fact, the van der Waals ZEKE spectrum in Fig. 24.8 demonstrates this effect for other levels. The ZEKE states here have been pumped via a whole series of low-frequency van der Waals modes as resonant intermediates in the S_1 state. The top spectrum is from the naked ground state and reveals the ionization potential; also the first few assign-

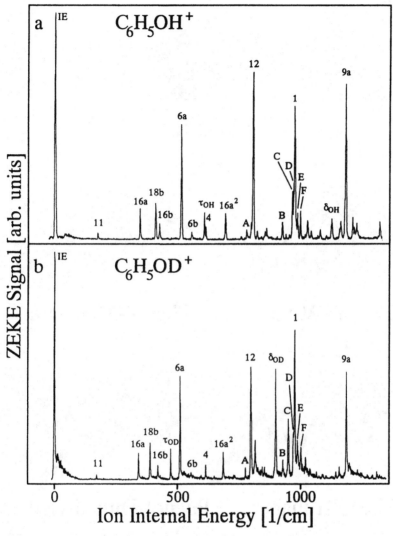

Fig. 24.5 ZEKE spectra of phenol via the S$_1$ 0^0 state.

able modes such as η enter. We have then one mode of the ξ_1 excited in the S$_1$ state, which gives a totally different spectrum again; excitation of one of the ξ_2 modes, one of the ξ_3 modes and an overtone of the $\xi_1 - \xi_3$ modes again leads to different spectra. This, of course, allows us to assign the states above and below these levels by making use of the Franck–Condon principle.

Phenol

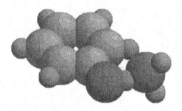

Phenol-Water   Phenol-Methanol

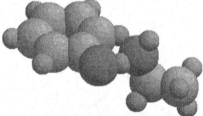

Phenol-Ethanol   Phenol-Dimethylether

Fig. 24.6 A space-filling model of the complexes investigated. Structures are only
approximate[50].

The new intermolecular coupling is seen in Fig. 24.9, in which this leads
to a harmonic out-of-plane wagging mode, but also to a very anharmonic
in-plane wagging mode. The in-plane wagging mode is a very anharmonic
type of mode as shown by the calculations by Hobza *et al.*[231]. This anharmonic wagging mode is predestined to interact strongly with the attached
water molecule. We have here a coupling between these two motions which

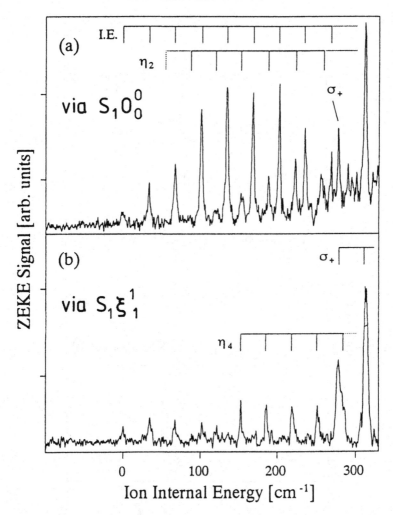

Fig. 24.7 The phenol–methanol complex near the IP[50]: (a) excitation originating from the vibrationless intermediate $S_1 0^0$ state and (b) the same, but now the origin is the first excited ξ_1 state of the intermolecular vibration in the S_1 state.

are out-of-phase motions, but, since they are of the same kind, it leads to a coupling of a similar motion. Overall, this results in a large spectral shift, which is observed in the spectrum.

To understand the spectroscopy of these phenol–water complexes, let us first review previously reported spectra of these systems (Fig. 24.10). At the top left of Fig. 24.10, the photoelectron spectra due to Fuke et al.[232] are shown. These are the photoelectron spectra of phenol, phenol–water and

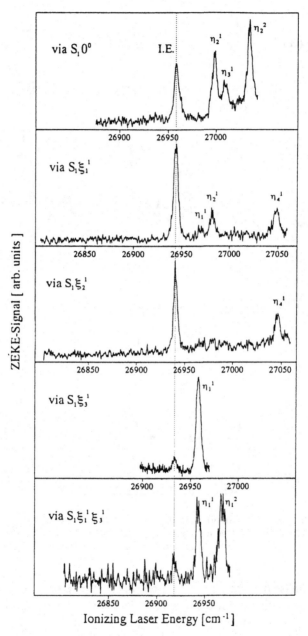

Fig. 24.8 Phenol–ethanol ZEKE spectroscopy[50] via various intermediate states in S_1.

Anharmonic Potential

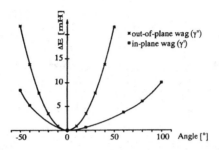

Coupling Between
Inter- and Intramolecular Modes

σ   18b

Fig. 24.9 Mode coupling as a result of strong anharmonicity along the intramolecular in-plane wagging modes, which are similar motions in phenol and in water[50].

several water molecules with phenol. They show onsets and some structure. If we measure the photoelectron ionization efficiency (PIE) curve for the phenol–water complex, as seen at the top right, the typical step-like functions with clearly defined initial onsets are seen. Below this is the ZEKE spectrum. Again, this measures the current with wavelength like the PIE spectrum, although most of the electrons are rejected – only the molecular energy levels are selected with a ZEKE detector and hence a different spectrum is produced. This results in far better resolution and indeed most of the stretching, wagging modes etc. can be observed and many can be assigned at least tentatively. This is seen at the bottom right as an excerpt of the first 250 cm^{-1} of the low-resolution ZEKE spectrum.

Not only can we start to assign these new intermolecular modes, but also we discover a shift in intramolecular modes. Within the strongly bonded molecular constituents, they shift considerably as a result of complex formation. A shifting in the molecular vibration due to the bonding with

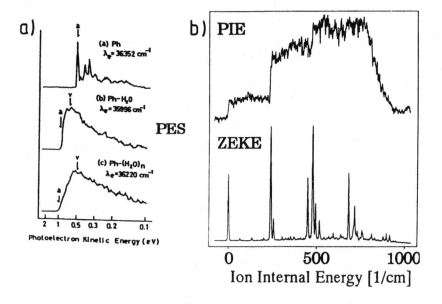

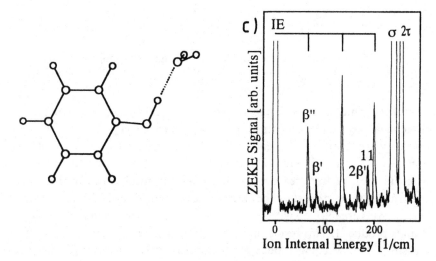

Fig. 24.10 The photoelectron spectrum of the phenol–water complex is shown in the middle of (a)[232]. In (b) the photo-ionization efficiency is compared with the ZEKE spectrum. In (c) the high-resolution portion of the ZEKE spectrum is shown.

Table 24.1. *Intramolecular phenol-localized vibrations (frequencies in* cm^{-1} *)*

	Experiment		Ab initio HF/3–21G*(O)	
Vibration	PhOH–H$_2$O (PhOD–D$_2$O)	PhOH (PhOD)	PhOH–H$_2$O (PhOD–D$_2$O)	PhOH (PhOD)
11 (a″)	189	177 (175)	218 (193)	184 (180)
16a (a″)	354	348 (344)	381 (392)	376 (364)
16b (a″)	435	428 (422)	458 (457)	442 (421)
18b (a′)	450 (435)	411 (390)	476 (451)	403 (382)
6a (a′)	516 (517)	514 (511)	520 (515)	516 (511)
6b (a′)	568	557 (555)	575 (574)	572 (571)
4 (a″)	636	616 (614)	656 (656)	643 (643)
12 (a′)	812 (804)	808 (799)	777 (773)	775 (768)
1 (a′)	977	977 (976)	875 (875)	870 (870)
τ_{OH} (a″)		608 (473)	1105 (781)	566 (457)

the water occurs, even within the phenol itself, but not for all modes (Table 24.1). For the ν_6 vibration, the shift due to the water is negligible, as is that for the deuterated species. In the case of the ν_{18b} mode, a carbon–oxygen bending vibration, a 10% shift due to complex formation is seen. Some of the vibrations themselves are quite small, being of similar magnitude to the van der Waals vibrations themselves. Thus shifts due to deuteration, complexation and other effects are present. Note that, usually, frequencies decrease with complexation, but sometimes, as for ν_{16b}, they increase due to the additional force of the hydrogen bond.

We can now compare these with *ab initio* calculations on a simple SCF level by Hobza *et al.*[231]. These are already very successful in being able to tell us something about the vibrational assignments. As a first example, ν_6, which had only a small shift, agrees well with the *ab initio* calculations. This correspondence is typical in my experience. The *ab initio* calculations have become important because, although by no means perfect, they are essential for a first assignment of these spectra. To make assignments, they must be reasonably accurate, but not necessarily exact. It should be pointed out that, without theory, these data are simply spectral lines without any hope of assignment, there being no transferable force fields available for a normal coordinate treatment of these intermolecular modes.

Our initial attempt to assign these spectra involved the use of molecular modelling methods. There are many elegant programs for this purpose. We

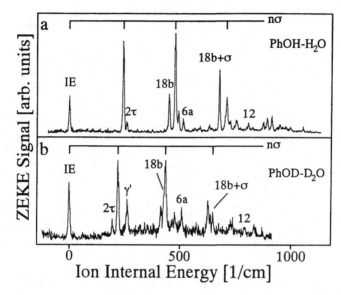

Fig. 24.11 Phenol–water ZEKE spectra. IE: 64027 (64020) ± 4 cm^{-1}. Intermolecular modes (values for H$_3$ isotopomer, values for D$_3$ isotopomer in brackets): σ 240 (221) cm^{-1}, γ 328 (264) cm^{-1} and 2τ 257 (197) cm^{-1}. Intramolecular modes: 6a 516 (517) cm^{-1}, 18b 450 (435) cm^{-1} and 12 812 (804) cm^{-1}.

attempted to assign all these observed lines to the various vibrations of the complex using such molecular modelling programs. For molecules, this is easy, but, in the case of van der Waals vibrations, it became impossible. Modes were hundreds of wavenumbers wide of the mark, making even a crude assignment impossible.

With the *ab initio* program, results from calculations are close enough to the observed frequencies for one to venture an assignment. In the phenol–water spectrum (Fig. 24.11) a large shift is observed in the torsional motion from the 2τ in the protonated to the 2τ in the deuterated species, a motion due to a proton moving. The overtone motion of the stretch, σ, beginning at the origin as a sequence, is a typical result for these systems. We see the intermolecular modes either directly or as a combination band on the intramolecular modes.

These are the first of the six new modes, specifically the stretching mode, the wagging mode and the torsion mode. What these modes are labelled is always a question of nomenclature. To define the motions, the actual displacements of the normal coordinate treatment produced in the calculation must be examined. The major problem was one of the assignments to which I have alluded previously. In Table 24.2, the case of the phenol–water cation

Table 24.2. *Phenol-water cation intermolecular vibrational frequencies* (cm^{-1})

Vibration	Experiment h_3/d_3	Ab initio h_3/d_3
Out-of-plane bend	67/64 (?)	87/82
β''	(1.05)	(1.06)
In-plane bending	84/–	124/117
β'	(–)	(1.06)
Torsion	$\approx 130/\approx 100$	202/163
τ	(≈ 1.3)	(1.24)
Stretching	240/221	275/259
σ	(1.09)	(1.06)
In-plane wagging	328/264	358/259
γ'	(1.24)	(1.38)
Out-of-plane wagging	–/–	451/330
γ''	(–)	(1.37)

is presented. Five of the six intermolecular vibrations for this cation can be seen, although the γ'' motion is not yet observed. The γ'' frequency is very large, even larger than that of the stretching mode. In Table 24.2, the six calculated intermolecular modes of the phenol–water system are listed. The calculations are not quite as nice. This is a restricted open-shell Hartree–Fock calculation and we have an ambiguity for two of the vibrations. For all modes we have a prediction and, therefore, even information on the basis of which to make the assignment. Although this is an excellent case even for simple *ab initio* programs, it is an unfortunate failure of molecular modelling of these intermolecular modes. Even sophisticated molecular modelling calculations were without success. This may be a special problem of weak complexes since we found such calculations to be highly useful in the case of strongly bonded systems for the interpretation of normal molecular spectra.

In review, the problem is the following. If you simply have a series of observed spectral ZEKE lines of a cluster you have absolutely no way to assign them without an *ab initio* theoretical calculation by which to identify the bend versus the torsion. The entire assignment would have been impossible without *ab initio* calculations. Here one has the beautiful situation in which theory is not just confirming a known experimental result, but indeed the experimental interpretation would have been impossible without it. The numbers, of course, will vary slightly depending on the calculation. They will also vary a little depending on which of the available quantum

Table 24.3. *Ionic intermolecular vibrational frequencies (cm^{-1}) for the phenol–methanol and phenol–water clusters*[50]

| Vibration[a] | [Phenol–methanol] | | [Phenol–water] | |
	Theory[b]	Experiments	Theory[b]	Experiment
Out-of-plane bending β''	51	34	73	67
In-plane bending β'	67	52	122	84
Torsion τ	102	76	180	≈ 130
In-plane wagging γ'	137/154	153	425	328
Out-of-plane wagging γ''	137/154	158	449	320
Stretching σ	310	278	275	240

Notes:
[a] Nomenclature adopted from the phenol-water case.
[b] Harmonic intermolecular frequencies (HF/3–21G*(O)).

calculational packages one employs, CADPAC[233] or Gaussian[234]. In one particular example of the phenol–water complex, the case of the out-of-plane bending mode β'', the result increases slightly from 73 cm^{-1} (Table 24.3) to 87 cm^{-1}. For all intents and purposes, these agreements are perfectly adequate for the assignments of these soft vibrations. The correspondences are in many cases surprisingly close and certainly adequate for assignment.

Consider another example, the complex of phenol bound to methanol as shown in Table 24.3. The σ stretching frequency is larger than that in the case of the cluster with water because the force constant is larger. In fact, σ for methanol has the largest value for all adducts tested, which is in agreement with the theoretical calculations. Note also the large out-of-plane wagging γ' for water in comparison with those of methanol and other adducts (Fig. 24.12).

In Table 24.4, a comparative list for five different phenol complexes (including phenol) that we have measured is presented. We have seen all six modes for three of the complexes. The σ stretching vibrations are again

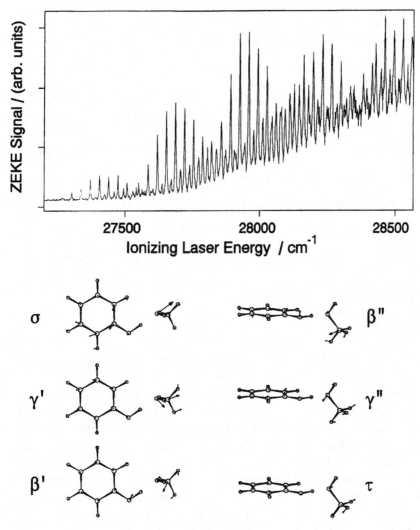

Fig. 24.12 The phenol–methanol complex. The intermolecular normal modes are
shown in the lower part.

among the largest, except for the case of water. The stretching vibration
behaves the way you think a stretching vibration ought to in that it has the
largest effect. Water seems to be an unusual case in which the waggings γ'
and γ'' are the largest contributors. The ionization potentials also are red-
shifted as one would expect, as are the origins of the S_1 states. These are
typical numbers which are displayed and are normal for the shifts as well
as for the origins of the S_1 state. Interestingly, the shifts in these S_1 origins

Table 24.4. *Spectroscopic data of several phenol–X complexes (X=water, methanol, dimethylether, ethanol and phenol)*

Species	IE (cm^{-1})	ΔIE (cm^{-1})	ΔS_1 (cm^{-1})	$\sigma(S_1)$ (cm^{-1})	Observed intermolecular ion frequencies (cm^{-1})
Phenol	68628	0	0	–	–
Phenol–H_2O	64027	4601	353	156	67, 84, **240**[b], 257[a], 328
Phenol–CH_3OH	63207	5421	416	176	34, 52, 76, 153,158, **278**[b]
Phenol–CH_3OCH_3	62604	6024	455	141	24, 57, 109, 141[c], 196, **275**[b]
Phenol–C_2H_5OH	62901	5727	410	162	25, 38, 53, 107, 248, **279**[b]
Phenol–phenol	63649	4980	305	120	19, **181**[b]

Notes:
[a] The first torsional overtone (2τ).
[b] σ stretching vibration.
[c] Uncertain assignment.

are larger than just the σ stretching vibrations in S_1 – which, of course, are less than the values in the ions, due to increased bonding.

Obviously, there are various effects present here, including mass effects and force-constant effects. In addition, some Dushinsky rotation in the excited state (re-mixing of several normal coordinates) whereby different states are being projected onto the ground state also is expected. A strong Dushinsky effect has been observed in the spectrum of benzene, which was the anomaly that caused so much trouble with the v_{14} and v_{15} vibration mixing in the normal modes of the S_1 state. The structure for the phenol–water system seen in Fig. 24.13 is that which comes from a CADPAC calculation. In the neutral species, the water–phenol distance (d in IV b 10) is 1.9 Å whereas in the ion it is somewhat shorter at 1.6 Å. The binding energy of the neutral species indeed is approximately 2000 cm^{-1} and that in the ion about 6000–7000 cm^{-1}. Again, this is a very low-level calculation, in particular a Hartree–Fock 6–31G* calculation. The six various types of intermolecular modes for the phenol–water system are found. Indeed, this now defines the nomenclature of these motions. Note that we have two symmetry blocks: the A″ modes which are symmetry-forbidden and the allowed A′ modes. However, all are seen, no doubt owing to channel coupling effects. This might also provide some hints for methods

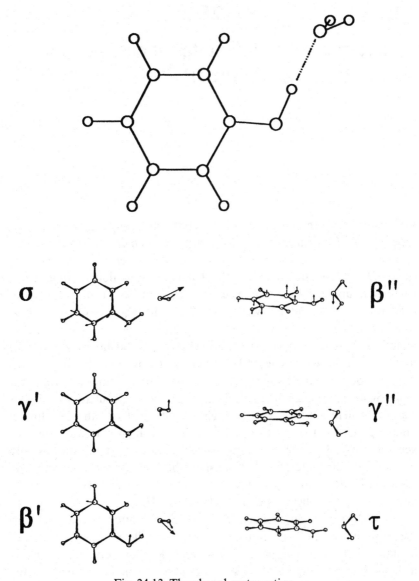

Fig. 24.13 The phenol–water cation.

to uncover difficult modes. It should be remarked that these are first attempts at assigning these important soft-modes here. Further methods, such as depletion spectroscopy, will no doubt lead to further progress in this field.

25

Lifetimes of ZEKE states at energies above the IP

One of the most exciting aspects of ZEKE spectroscopy is the extreme lifetime of the ZEKE states, which for molecules even extends to states substantially above the ionization potential. Reiser *et al.*[35] had initially observed that the states just below the IP have an anomalous lifetime, which can extend up to 100 μs. What is even more surprising is that such states exist not only near, but also well above, the IP, having been observed[136] up to an excess energy of 8 eV. These are states of high n in their respective manifolds, hugging the ion state from below with total energy sufficient to ionize, but which remain neutral on a very long time scale. In other words, it is not correct that all states within the ionization continuum will autoionize with sub-picosecond rates. This longevity is a peculiarity of ZEKE states that here are formed from high-n Rydberg states, an effect that is unique to molecules and that can even lead to high core excitation. In general, molecular states are rich in states in the continuum whereas only very few atomic state manifolds exist in the ionization continuum. The situation is shown in Fig. 25.1. Atomic state measurements are typically achieved by pulsed-field experiments below the IP.

The molecule is characterized by core state excitations above the IP on the right-hand side of Fig. 25.1. The $v_1^+=1$ state will be the first vibrational excitation in the ion. This state has a separate Rydberg series leading up to it. Upon just reaching the $v_1^+=1$ state from below, one finds 'hugging' this state a very-long-lived ZEKE state in a narrow band of Rydberg states (near $n=150$), which become long-lived ZEKE states. This $n=150$ state converging to the continuum at $v_1^+=1$ survives 50–100 μs, much like the state below the molecular IP, except that now ionization is not an open kinetic channel although it is open energetically. ZEKE states, even above the IP, do not avail themselves of this open energetic channel. This is also true for $v_2^+=1$ at $n=150$ and $v_3^+=1$ at $n=150$, etc. This gives rise to the ZEKE

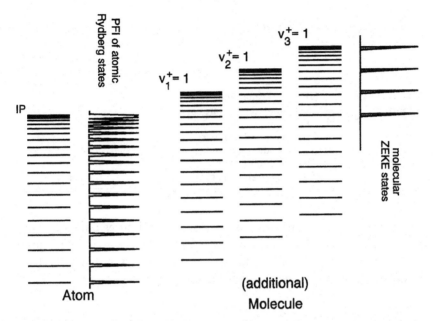

Fig. 25.1 Channel coupling to produce long-lived vibronic molecular ZEKE states above the IP. In comparison, pulsed-field-ionized atomic levels are normally below the IP.

spectrum for cations. Although one might have expected such long-lived isolated states near the IP, it is interesting and surprising to note that they exist with high excess energy even well above the IP.

One can ask the interesting question of how much excess energy can be stored in this ZEKE state before the electron is ejected on its own and the ZEKE state is destroyed. This would provide a natural upper limit. So far, ZEKE states have been observed by Hepburn *et al.*[136] even for the third electronic state of NO^+ some 8 eV above the IP (Table 25.1). Similar results have been seen for HBr and benzene. At present, there is a major effort with synchrotron radiation to go to still higher energies, using the ZEKE technique[39]. High-resolution synchrotron experiments are here predicted to be a rich field. Note that ZEKE resolution is not energy-dependent, whereas electron-monochromator resolution is.

Normal photo-ionization is quite different. In single-photon or electron ionization, one populates a state (Fig. 25.2) that then undergoes strong dissociation. In multiphoton ionization, one sequentially adds more photons until the successive thresholds to fragmentation have been reached, described by a ladder switching mechanism (Fig. 25.2, Fig. 25.3). In the

Table 25.1. *Excited ZEKE states*

	System	Energy above threshold (eV)
Müller-Dethlefs	NH_3^+, umbrella mode ($v^+=9$)	1
Hepburn	Electronic excited states of NO^+	8
Knee, Neusser	Dissociation of van der Waals modes	0.1
Hepburn	Dissociation of HBr^+	≈ 2
Selzle	Fragmentation of benzene$^+$	>4.4

case of ZEKE, excitation is to a neutral state that ionizes only after the application of a field. The longevity of these ZEKE states is, of course, just due to the high-ℓ, m_ℓ state keeping the electron away from the core, thus turning off the perturbations responsible for decay.

25.1 The onset of ionization

In fact, even the point of ionization at the IP is problematical. If, as is typical, a slight electric field is applied while scanning the laser to shorter wavelengths, the onset of ions at the IP will eventually be obtained. This measurement leads to the photo-ionization efficiency (PIE) curve. As we now know, even this statement is problematical. It is correct that one sees the onset of ions, but it is not the ionization potential that is measured. Rather, the total current will reflect ZEKE states below the IP that produce ions in the presence of the field and that produce a false onset for the IP. At the top of Fig. 25.4, the onset curve for ions is presented. This is a false onset. With a delayed electric pulse, the ZEKE states can be located and these states account for the initial rise of the curve. One must subtract this ZEKE signal from the original total current signal to obtain the curve below, which now displays the correct onset of ionization. In other words, the total ion signal must always be corrected by subtracting the ZEKE signal.

In Fig. 25.1 one sees that here $v_1^+=1$ produces ZEKE states. On applying this principle to benzene, we observe that $v_6^+=1$ in the ion produces a long-lived ZEKE state with a lifetime of some 20–100 μs. One could measure the decay of these ZEKE states, which will no doubt be multi-exponential, but a lifetime of 30 μs (Fig. 25.5) can be estimated. If one increases the energy in the benzene ion substantially, one gets decomposition, leading to a normal mass spectrum of the molecular ion.

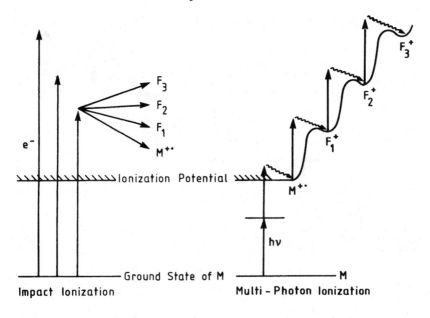

e⁻ - IMPACT   1-STEP PROCESS

    change of e⁻ intensity   no change of mass spectrum
                                change of ion yield
    change of e⁻ energy   minor change of mass spectrum
                                change of ion yield

MUPI   MULTI-STEP PROCESS

    change of hv intensity   drastic change of mass spectrum
                                minor change of ion yield
    change of wavelength   change of mass spectrum
                                different fragmentations
                                change in ion yield

Fig. 25.2 Electron impact versus multiphoton ionization.

25.2 Core photodissociation

One of the curious further properties of ZEKE states is that it is possible
to dissociate the core without affecting the ZEKE electron directly. This is
the extreme example of the electron acting as a spectator with respect to the
positive core; it is even essentially ignoring not only excitation, but even the
far more violent photodecomposition of the core.

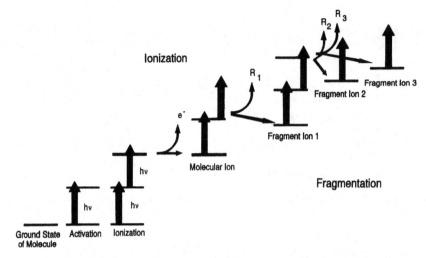

Fig. 25.3 The multiphoton-ionization ladder switching model.

Scherzer *et al.*[124,125] and Neusser and Krause[27] performed ZEKE core photodissociation at high energies and actually observed dissociation in multiphoton-pumped ZEKE states. Photodissociation of ZEKE states of benzene leaves the ZEKE electron untouched, dissociating only the ionic core. The ZEKE electron is now attached to a product state and a neutral fragment is ejected. The spectator electron at very high n is now encircling the product states. At this energy, we see the core bursting apart but the ZEKE electron is not affected by this process.

25.2.1 The dissociation of the argon complex

The simplest case to consider is the benzene–argon complex. This is shown in Fig. 25.6. After photoexcitation we have a vibrationally excited state of the ion of the complex with its Rydberg electron in a spectator state. Now the argon departs as a neutral species, leaving the benzene in a Rydberg state. The interesting conclusion is that this violent process of shaking up the core does not cause the electron to depart.

That argon leaves the core of the ZEKE state of the benzene–argon complex is perhaps understandable. This is an interesting point of fragmentation without influencing these high-ℓ, m_ℓ electrons undisturbed in their planetary orbit. This opens up a whole new way of studying doubly, or more highly, ionized molecules by further excitation from a ZEKE state. Analogous ionic core excitation (ICE) has been done for high n in atoms[235].

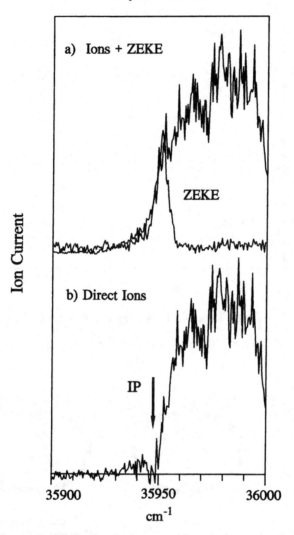

Fig. 25.4 The onset of ionization for benzene. Above is shown the total creation of ions. The ZEKE intensity is seen to be the leading edge of this signal. These states are here slightly below the IP so that they should not be counted in the measurement of the IP. Subtracting this ZEKE peak from the total current gives the spectrum below with the correct onset of the IP. Hence, technically the rise of the total current is not quite the correct IP.

These ZEKE systems represent a new kind of system for molecules that can be constructed also with artificial Rydberg states that are not prepared optically. This method represents a new kind of Rydberg mass spectrometry. This is an interesting game because, by microscopic reversibility, one

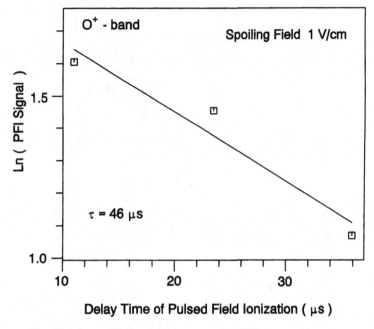

Fig. 25.5 Long-lifetime components of ZEKE states of benzene.

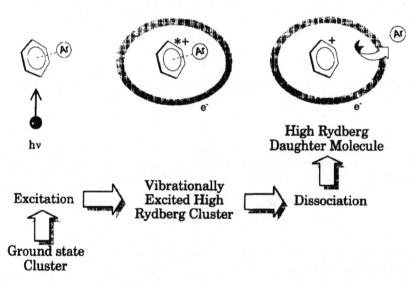

Fig. 25.6 Dissociation of highly excited Rydberg clusters[140]. Excitation of the ZEKE state also leads to photodissociation of the core without loss of the electron.

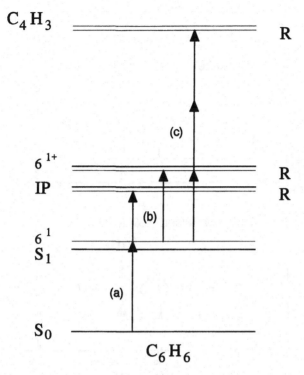

Fig. 25.7 Energy regimes for ZEKE states: (a) excitation of ZEKE states just below the IP, (b) excitation of ZEKE states just below ν_6^+ above the IP and (c) further excitation of the states in (b) in a multiphoton mode that leads to fragmentation of the ZEKE core without ionization.

can in principle produce these states in the opposite direction even at almost arbitrary energies. This has been confirmed experimentally in our laboratory[122] (see the next chapter).

25.2.2 The photodissociation of the benzene core

The interesting thing, of course, is that you have a very curious mechanism now. This I will pictorialize for the excitation of benzene, in which case it is even more surprising. In this case, we start with excitation to the $\nu_6^+=1$ ZEKE state, but add a second laser of substantially higher intensity. Again, this is done in a zero-field environment with a small spoiling field to remove all ions formed as a result of direct photoexcitation. The draw-out field can again be delayed. The question of whether any ZEKE states have been formed can be asked (Fig. 25.7). Much to our surprise a substantial ZEKE

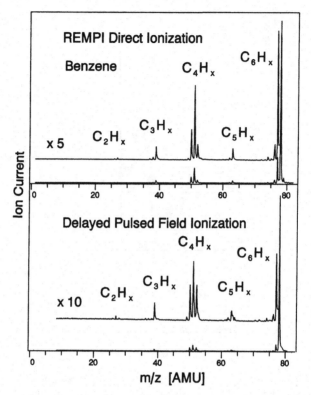

Fig. 25.8 A comparison of a REMPI mass spectrum and a ZEKE mass spectrum.

signal surviving many microseconds is observed and the pulsed-field analysis reveals these ZEKE states to be fragments, such as C_4H_3. This indicates that benzene is photodissociated in the core without touching the orbiting ZEKE electron. After photodissociation, a pulsed field identifies the process as a ZEKE state producing an ion that appears after the pulsed field. We therefore have found a further property of ZEKE states, namely, that these states have orbiting electrons in high-ℓ states that prevent them from auto-ionizing. Furthermore, they are oblivious to fracturing of the core. The normal photodecomposition of ions in a REMPI mass spectrometer, which is always done with a small field in normal mass spectrometry, is shown in Fig. 25.8 (top spectrum). Below is a ZEKE spectrum with mass analysis after a pulsed delayed field. This demonstrates that even highly excited, photofragmented states survive without losing the ZEKE electron. The rough similarity is expected since the ion is similar in the two experiments. Nevertheless, we want to demonstrate that all these states are

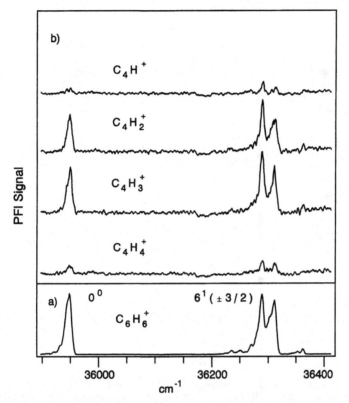

Fig. 25.9 Delayed PFI spectra of benzene and C_4H_n fragments. This demonstrates the ν_6^+ origin of the decomposed ZEKE states.

pumped up from the initial $\nu_6^+=1$ (Fig. 25.9) and are ZEKE states. The bottom curve shows the ZEKE spectrum of the origin and the $\nu_6^+=1$. The top four cases show various fragments. Clearly these fragments are only produced at energies at which there is prior absorption by a ZEKE state at the origin or at $\nu_6^+=1$. Hence, these are up-pumped ZEKE states that then produce fragmentation – without disrupting the spectator electron of the ZEKE state.

The apparent independence of the ZEKE electron from extreme internal perturbations of the core, even though the slightest field would have affected it, is of great interest. We have seen that this spectator state explains the abnormal lifetime of the state. We have also seen that this spectator electron is immune to photodecomposition of the core. We see in this photoexcitation that there are no resonance requirements (similar to the cuckoo effect, q.v.), thereby showing that ZEKE states can be produced as orbiting

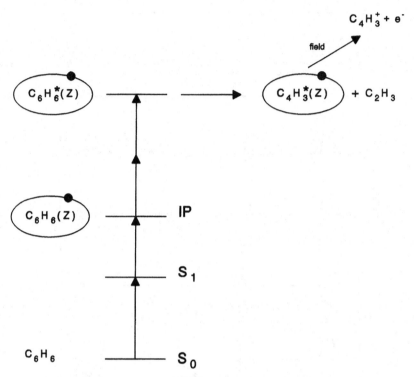

Fig. 25.10 Dissociation of highly excited Rydberg molecules. This is excitation of benzene via the ν_6^+ state of the ion, with additional multiphoton dissociation leading to dissociation in the core of the ZEKE state without losing the ZEKE electron. The electron is ejected for test purposes in a later pulsed field.

collisions without the optical resonance requirements of normal ZEKE spectroscopy. Its lifetime is 50–100 μs. This leads one to the question of what affects the lifetime. Black-body effects on this time scale can probably be excluded. It is probably a decay into other electronic states. One could ask whether changing from C_6H_6 to C_6D_6 affects the measured lifetimes and check the vibrational state dependences. Detailed results have now shown[229] that the 75 μs lifetimes show no isotope effects – hence precluding vibrational effects here – a not entirely expected result.

Thus, one sees a new property of ZEKE states. The transformation of a Rydberg state into a high-ℓ, m_ℓ ZEKE state has moved the electron to an orbit in which it appears to become a spectator to the system. It does not substantially interact with the core, so the electron does not depart. Furthermore, it appears to be so uncoupled from the core that the core can be photodissociated without the electron departing (Fig. 25.10). The electron now belongs to a ZEKE fragment and a ZEKE state for the fragment has been produced. It should be noted that an arbitrary amount of high-intensity energy has been introduced to photodissociate the core and to produce a proper ZEKE state of the core. The departing fragment has not dislodged the distant electron. This will become an important property of ZEKE states that separates them from normal Rydberg states.

One of the most important aspects of this ZEKE mass spectrometry is that its resolution is much better than that of a REMPI spectrum. Notice that we again readily see the Jahn–Teller splitting of the degeneracy in the v_6^+ vibration in Fig. 25.9 in the spectrum. This would not be observed in a REMPI spectrum. Note also that a REMPI mass spectrometer is simply converted into a ZEKE mass spectrometer by waiting with the draw-out field until the laser has been switched off and eliminating the direct ions.

26

Effects of ions on ZEKE spectra

This, of course, raises some very important questions. Are the ions produced from up-pumped direct states? Or are they produced by delayed ionization of the neutral states as we say here? This has been answered by ionization delay. We could contemplate a further possible artefact of the situation whereby one could obtain these results by producing a high-concentration mixture of ions and neutral species. ZEKE electrons are always sitting in proximity to a large background of ions and the ion could simply act like a huge vacuum cleaner picking up all the ZEKE electrons that are produced, regardless of where they are produced. This then would produce false ZEKE states. In fact, this would be a bimolecular type of ZEKE state production and not a unimolecular production of a Rydberg state migrating to become a high-ℓ ZEKE state. This question has been answered by a sophisticated three-laser experiment[122].

Our experiment involves a three-laser excitation scheme with two different isotopic molecules (Fig. 26.1). For a worst-case test, we choose the two isotopomers, light and heavy benzene, C_6H_6 and C_6D_6, and utilize three lasers. Laser number one operates at the ν_6 resonant energy of C_6H_6 (Fig. 26.2) while laser two operates at the ν_6 resonant energy of C_6D_6, both in the S_1 state. The third laser will produce the possible ZEKE states in both systems. Radiation from the third laser can be absorbed by the ZEKE states either in light or in heavy benzene, starting from their respective ν_6 transitions. As we increase the energy of the third laser above the IP of either species, one produces not only ZEKE states, but also a great many more direct ions. These could in principle always act as vacuum cleaners for stray electrons. Very slow ZEKE electrons, as they are formed, should be particularly good candidates for this vacuum cleaner. Hence, the first ZEKE electron formed would become a candidate for a crossover to one of the ions of the other species. ZEKE electrons from one of the transitions of C_6H_6

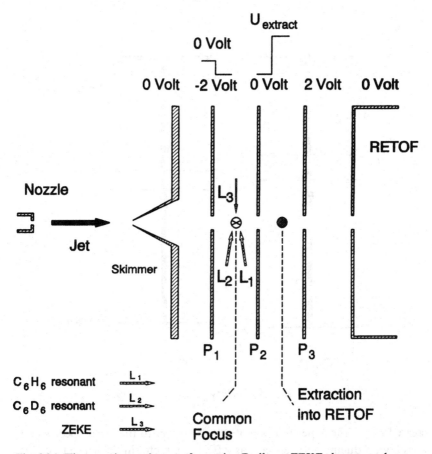

Fig. 26.1 The experimental set-up for testing Rydberg–ZEKE electron exchange.

now would jump over to this vacuum cleaner of C_6D_6, possibly producing a false ZEKE state. Although many electrons of varying energies are always produced, the slow ZEKE electrons (at their threshold of production) should be particularly good candidates for crossover. If all electrons were candidates, this would be a plasma effect. From the foreign ZEKE electron we would have a false state in the sense that it is a bimolecular ZEKE state undergoing detection as an ion for a species other than the species produced directly by a photon. Since the transfer is particularly effective for a ZEKE electron, it would produce a false state at a well-defined energy. It would be false in the sense that scanning the laser with a delayed field and the mass selector set on the mass of $C_6D_6^+$ is producing a conventional C_6D_6 ZEKE spectrum, with false peaks at the energy of the $C_6H_6^+$ state. These would be

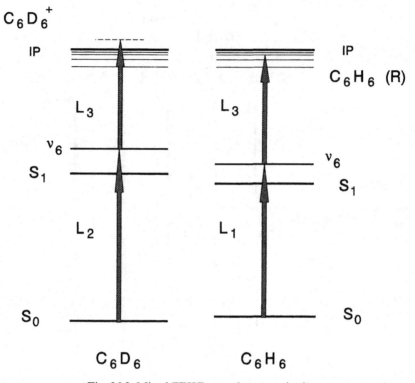

Fig. 26.2 Mixed ZEKE two-photon excitation.

created by an interloper popping its electron over to the other species and possibly producing a state that has ZEKE character in the sense that it is picked up in a ZEKE spectrum at the mass of $C_6D_6^+$ but is not produced from an optical Rydberg transition of C_6D_6. The C_6H_6 produces the ZEKE electron much like a cuckoo and puts it in the nest of the C_6D_6 to give birth to a pseudo-C_6D_6 ZEKE state where in general no native C_6D_6 would have had a ZEKE state (Fig. 26.3).

An actual experiment that uses these two species C_6H_6 and C_6D_6 to follow the process can be considered. We look first of all at the C_6D_6 mass. In this particular case, we scan the spectrum and observe the normal onset and the ν_6^+ of the C_6D_6 ZEKE spectrum. We focus on the detection only for the heavy mass and we adjust the mass in the mass spectrometer so it measures only the heavy mass at the top of Fig. 26.3. We see exactly the 0–0, a little of the ν_{16} and a ν_6 splitting here and that is all. Now laser number one is turned on to produce ZEKE states from C_6H_6 also (Fig. 26.3). Now the ZEKEs are produced from both species and when we run the same C_6D_6

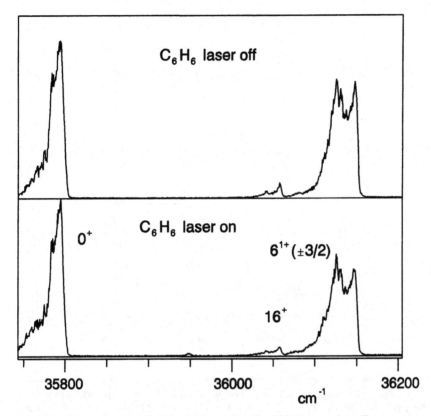

Fig. 26.3 Delayed pulsed-field ionization of C_6D_6. This shows virtually no cross-talk between the two isotopomers in a normal ZEKE spectrum since the two spectra are virtually identical.

spectrum (bottom curve) one can see that the C_6D_6 spectrum looks almost the same even though $C_6H_6^+$ ions are now also present in the background, but not detected due to mass selection. The C_6H_6 spectrum (not shown here) has a ZEKE threshold between the peaks. If one looks at the C_6D_6 ZEKE spectrum in the presence of $C_6H_6^+$ ions the spectrum is virtually identical to the pure C_6D_6 ZEKE spectrum. There is a little peak in the middle. The question we can ask ourselves is the following: is that blip real or not? By amplification, the blip can be increased by a factor of 30 (Fig. 26.4) and, to our great surprise, this blip is indeed at the ZEKE threshold of C_6H_6 (top spectrum). We show that this peak (though present only to about 2% intensity) represents a novel and very curious phenomenon. A preponderance of $C_6D_6^+$ ions is always being produced by the direct process which will detect, as a vacuum cleaner, the ZEKE electrons from the C_6H_6

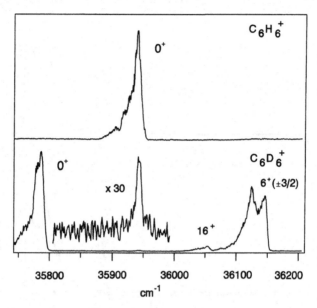

$$\text{Rydberg} \, (C_6 H_6)^{(0)} + C_6 D_6^+ \xrightarrow{\;e^-\; \text{Transfer}\;}$$

$$C_6 H_6^+ + \text{Rydberg} \, (C_6 D_6)^{(0)}$$

Fig. 26.4 Delayed pulsed-field ionization of C_6D_6. Closer examination reveals the cross-talk to be negligible, under 2%. This is shown in the lower spectrum, in which this peak is blown up to show the slight interference from the other isotopomer. Note that this peak is sharp and not simply a plasma effect.

ZEKE transition albeit it is a minor process. This makes a pseudo-state that then is measured at the wrong mass for $C_6D_6^+$, since a cuckoo electron is produced by C_6H_6 but hatched by C_6D_6. This false C_6D_6 ZEKE is made by borrowing the ZEKE from the light benzene and then plopping it over onto the $C_6D_6^+$ ions. Indeed, the effect is small, but it is present and observable. One can argue whether this is seriously interfering with normal ZEKE spectra. The effect measured is about 1 or 2% in the entire spectrum and clearly not a serious threat to direct excitation. Although this effect has to be watched in all cases, it is of minor import in most ZEKE experiments. One has to be careful because, as in all experiments, if one simply increases the laser intensity in an uncontrolled fashion, one will obtain ZEKE spectra that are meaningless, leading to many false interloper or cuckoo peaks. It is, of course, in general not surprising that experiments with uncontrolled intensity can lead to spurious results.

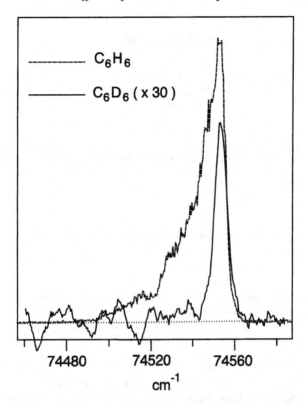

Fig. 26.5 ZEKE electron transfer. Here the slight cross-talk is blown up and compared with the originating C_6H_6 peak. Note that the transfer peak is narrower, showing that the transfer is favoured by high Rydberg states.

When this cuckoo peak is compared with the normal ZEKE peak in C_6H_6 a curious observation is made. Notice that the cuckoo peak is much, much narrower than the normal ZEKE peak (Fig. 26.5), which is an important aspect. This demonstrates that the cuckoo effect has a strong dependence on n, the quantum number of Rydberg levels attached to it. This indicates that the mechanism is most likely to involve electron transfer between two ion cores[236] at high n. In a kinetic description, this cuckoo effect arises from a collision of a ZEKE excited molecule M(R) with an ion N^+. When the two molecules approach, they form a complex that can be described as two ion cores with one common ZEKE electron and, on their flying apart, the ZEKE electron has a chance to stay with either part:

$$M(R)+N^+ \rightarrow (M(R)\,N^+) \rightarrow M(R)+N^+$$
$$\rightarrow M^++N(R)$$

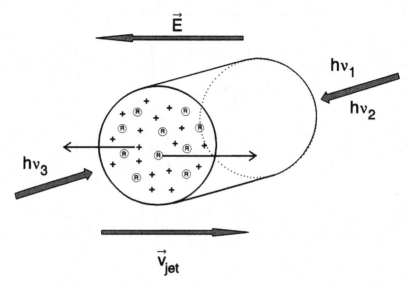

Fig. 26.6 Separation of ions and Rydberg states due to the spoiling field. This model of ions interpenetrating neutral species as a result of a 1 V cm^{-1} gradient producing cross-talk.

Clearly this mechanism will favour the highest Rydberg states, hence accounting for the sharpening of the peak and actually increasing the resolution. The important point to remember is that this transfer only occurs at the resonance energy of the cuckoo, not at all energies like in a plasma effect. In general the ions and electrons do not get together, the effect only works at threshold. When estimates of the ion and ZEKE densities in the beam are made, one arrives at a capture cross-section of about four Rydberg orbits at $n=150$ (Fig. 26.6). The cross-section for this process is indeed enormous[236] (Fig. 26.7, Fig. 26.8) and rises with n. The total falsification of a ZEKE spectrum as a result of this cuckoo effect, however, is negligible, at least under normal conditions. It is interesting and important to show that this cuckoo effect also works in the opposite direction for the other isotopomer, i.e. the pollution of the C_6H_6 ZEKE spectrum by ZEKE electrons from C_6D_6 (Fig. 26.9). Furthermore, it can be shown that this effect does not change whether or not a field is present at the time of photoexcitation (Fig. 26.10). With this three-laser experiment, it could also be shown that, in the photodecomposition of ZEKE states (Fig. 26.11) in which the ZEKE electron stays in the spectator orbit, the fragments produced have a normal unimolecular origin from the core and are not produced from this cuckoo effect. This could be proven by selecting fragment

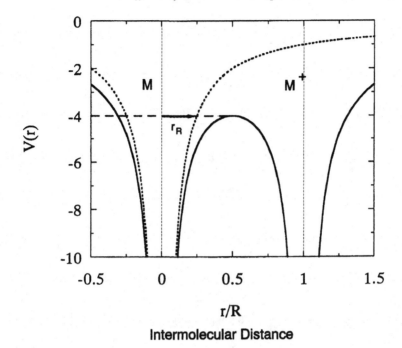

r/R

Intermolecular Distance

Fig. 26.7 The potential energy diagram for ZEKE electron exchange. This shows preferential tunnelling of high Rydberg states, leading to ZEKE electron pick-up by the other isotopomer.

masses that are unique to both molecules (Fig. 26.12). The cuckoo effect can be considered to be insignificant for a carefully recorded ZEKE mass spectrum (Fig. 26.11) but is of a particular interest on its own.

In summary, one can say that the cuckoo effect is present, but of negligible practical importance for normal ZEKE spectroscopy. It can be of importance in more complex systems and one always must bear it in mind. It will be of major significance for large-cluster spectroscopy.

An interesting application of this cuckoo effect has recently been given by Rednall and Softley[237], who utilized this effect by transferring the ZEKE electron to an added gas, SF_6. By now measuring SF_6^- one has a technique equivalent to electron detection of the ZEKE signal.

These results are also of interest in view of what has been said about the Rydberg states being like Swiss cheese; namely, that you can go through them. This is not the case. There are really gigantic resonances present. This cuckoo effect is a sharp, but false resonance and this is as it should be. It is important to point out, however, that it can be controlled and it will only affect normal ZEKE spectra in a very small way.

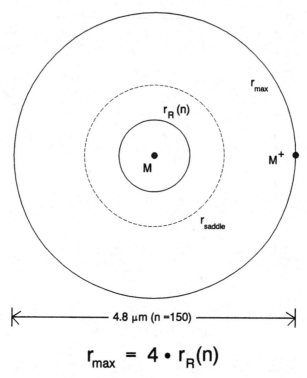

$$r_{max} = 4 \cdot r_R(n)$$

Fig. 26.8 The cross-section for ZEKE electron transfer.

The cuckoo effect leads to a very interesting conclusion regarding the nature of ZEKE states. We have already alluded to this for the case of decomposition in a ZEKE mass spectrum. If we look at the false ZEKE state in C_6D_6 produced between the two real states in Fig. 26.4 again, the origin and the v_6^+ ZEKE state can be observed. Note that, at 74 560 cm^{-1} (Fig. 26.5), there is only this false ZEKE state due to the cuckoo effect, specifically an ion of $C_6D_6^+$ capturing a ZEKE electron from C_6H_6. This only happens at the resonance of the C_6H_6 ZEKE state and, hence, the capture is only for resonant ZEKE electrons. It is important to note that this is not a general plasma effect. Although there is a large background of $C_6D_6^+$ ions and electrons from C_6H_6, only the ZEKE states of C_6H_6 produce the cuckoo effect in C_6D_6. The C_6D_6 is a properly behaved ZEKE state since it has a long lifetime and undergoes ZEKE detection, even though it is a false state in the sense that there is no optical Rydberg transition in C_6D_6 at this energy that could have produced a ZEKE state. It is real in the sense that it is, by all indications, a true ZEKE state. Hence, we come to the inter-

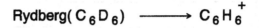

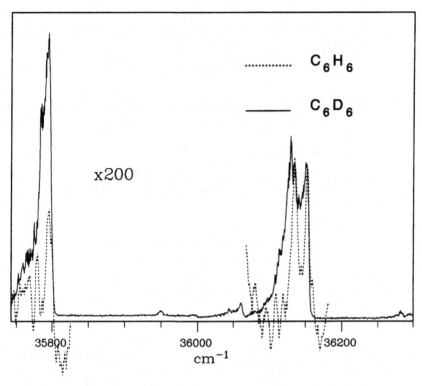

Fig. 26.9 ZEKE electron transfer. Transfer from C_6D_6 to C_6H_6, showing that this process can proceed from either isotopomer to the other[139].

esting conclusion that ZEKE states do not necessarily require optical Rydberg transitions in their early time history but have an existence of their own. This again indicates the unique character of these ZEKE states. In this case, it is more analogous to electron capture in an orbiting resonance than it is to a perturbed Rydberg state. It would also be an interesting resonance for Rydberg electron transfer. All this has important consequences for the possible field of ZEKE-state chemistry.

Hence ions in the background do not have a plasma effect, but they can via a minor process transfer a ZEKE electron to produce a false but nevertheless sharp cuckoo state in the spectrum. On the other hand, as proposed by Chupka[62] and as shown for xenon atoms, long-range ions affect ZEKE intensities. The effect therefore is present and has been shown also for benzene at low concentrations. Ions break the symmetry and thereby

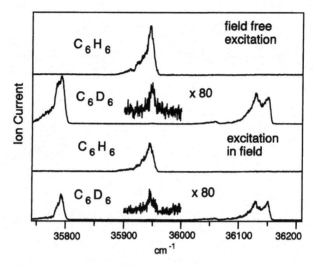

Fig. 26.10 ZEKE electron transfer. This demonstrates that the crossover can occur in pulsed delay experiments (top two plots) and for in-field excitation (bottom two plots).

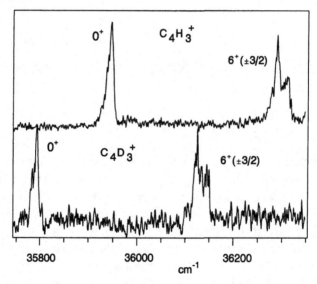

Fig. 26.11 ZEKE electron transfer. The small 2% peak corresponding to cross-talk between the two isotopomers is so small that it is not observable for fragments.

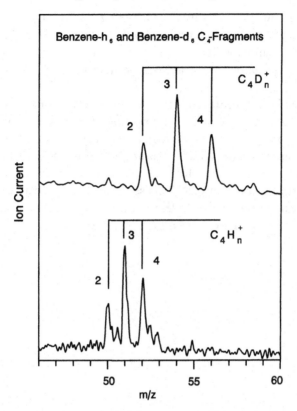

Fig. 26.12 Fragmentation of mixed isotopomers of benzene shows no mixed fragments, hence no cross-talk in fragment production.

induce m_ℓ transitions that produce ZEKE states. Further work indicated that this effect is also present for some diatomic molecules such as NO and I_2, although these experiments were femtosecond studies with typically enormous optical fields present. Our benzene experiments, on the other hand, showed the absence of this effect under normal experimental conditions[130]. If we extend our measurements to a typical low ion concentration we can force the system also here to produce a loss in signal.

The situation can be summarized as in Chapter 22 in terms of a unified mechanism in which photons produce low-ℓ Rydberg states. At values of $n > 100$ these are Stark-split by fields of some 10–20 mV cm^{-1}. These new eigenstates are no longer ℓ states but rather Stark states containing many ℓ components.

Added ions produce an additional effect in breaking the symmetry of the field as a result of the inhomogeneous component of the field produced by the ions. This now suddenly turns on m_ℓ mixing, thus driving the system to higher values of ℓ and m_ℓ. As such the Rydberg states are transformed into long-lived states since the electron now stays away from the ionic core in the manner of a spectator. These ion-transformed Rydberg states are here termed ZEKE states.

Still higher ion concentrations will widen the ZEKE bandwidth from some 8 cm^{-1} to some 15 cm^{-1}. This simply means that increasingly high n states can, if sufficient ions are present, also be converted to ZEKE states. The interesting experimental observation, however, is that this increased ion concentration does not affect the resolution one can obtain from a slicing experiment (Section 6.2). This seems to indicate that n-changing collisions are not yet operative under these experimental conditions. This is true even for long-beam experiments (Section 6.3).

27

Summary

The ZEKE method I have outlined can be categorized into several experimental methods (Table 27.1). At threshold, the ZEKE electrons observed are emitted without any excess energy and thus are detected after some delay with steradiancy analysis. This is the case of choice for anions in the absence of any bound states by stripping off the electron from a mass-analysed spectroscopy of neutral ground states. This would be interesting for mixtures (e.g. cluster beams) or short-lived intermediates with stable anion states. Just below threshold, at $n>100$ the ZEKE signal can be observed from the special bound states, which are very-high-n Rydberg states that differ from the optically populated states by having a particularly high orbital character and hence an anomalously long lifetime. Such thresholds were discovered for all states of the molecular ion even though they are at energies far above the IP as 'islands of stability' within the continuum. One can observe the ZEKE state from the corresponding positive ion. One could combine electrons and ions into a ZEKE coincidence technique. The latter has many additional possibilities but, of course, this is only just being realized owing to experimental difficulties. This latter method will be a very powerful new technique when the appropriate lasers or high-resolution synchrotron sources become available. It should also become a very important technique for the precise state selection of molecular ions for further experiments.

Quite apart from the details of the quantum states involved, we here show that ZEKE states already have an operational definition that distinguishes them from Rydberg states. In the first place ZEKE states typically evolve from optically populated Rydberg states, but only if $n>100$. Below these high-n states the Rydberg lifetimes are too short to allow for a facile transformation to ZEKE states. Under typical conditions, which are also the conditions of this experiment, this evolution of the high Rydberg

239

Table 27.1. *ZEKE methods*

Threshold detection
    directly formed ions at or nearly above threshold measured by
        a)   DC field present with photons
        b)   delayed field extraction and concomitant arrival time amplification

'Last state' below threshold
    Not ionized until a delayed field is applied
    Ionized by photons when field present

States in the ionization continuum above the IP as islands of stability

Fragment ZEKE
    Fragmentation of the ZEKE core

Multiple ionization from ZEKE states

Coincidence ZEKE
    State selection for molecular ions

Induced ZEKE
    ZEKE electron from another molecule captured by ion to produce false
    ZEKE state
    The cuckoo effect

---

states into ZEKE states operationally takes place only in the presence of ions and some stray field. A further criterion is that, at typical fields of 0.5–5 V cm^{-1} Rydberg states are operationally distinct from ZEKE states in that Rydberg states disappear as a result of these fields whereas ZEKE states are immune to such field effects (except for a minor additional effect of field ionization). Hence these are two effects, summed up in the unified mechanism of Held *et al.*[219], which operationally distinguishes Rydberg states from ZEKE states in these typical experiments.

We here represent the current state of the effect, reflecting only the beginnings of this method. We have attempted to show that this spectroscopy is useful for a number of molecular systems, though in this primer we have demonstrated it for only a few cases. We and many other groups have measured these effects and the reader is referred to our ZEKE home page on the web for an up-to-date bibliography[238]. We have here shown representative results for nitric oxide, ammonia (a case demonstrating nuclear spin separation), hydrogen sulphide (with spin–orbit coupling), benzene, *p*-difluorobenzene (with Jahn–Teller and Renner splitting) and a series of van der Waals complexes (for the first time, all six modes of these van der Waals intermolecular vibrations can be observed).

One could ask whether this spectroscopy can detect all possible transitions or whether it also is restricted by forbidden transitions, i.e. are there completely forbidden transitions? Our experience is that the peculiar physics of this technique permits one to see virtually every possible transition, albeit some weakly, i.e., every transition that is energetically possible. Both species of van der Waals modes with the allowed A' and forbidden A'' symmetry are involved in water–phenol complexes and both are seen. This is a direct result of strong channel coupling and not just weak perturbations due to auto-ionization. Light molecules form a class of molecules that typically observe the Franck–Condon principle, whereas Ag_2 is a representative molecule that strongly violates the Franck–Condon principle. In fact, it represents the extreme case of channel coupling to the ultimate end of all possible lateral transfers that are possible among the Rydberg stacks. This is to be expected since heavier systems are rotationally more congested. Clearly, there are many cases in between. Some cases show propensities with weaker forbidden transitions. In reality, their absorption follows optical selection rules, but they subsequently transfer to other ladders by channel coupling, showing up in places not vertically, but horizontally populated, thus apparently violating the Franck–Condon principle. In general one can say that, if the direct transition to a high-n Rydberg state is Franck–Condon-allowed and spectrally accessible, one sees the normal Franck–Condon-allowed ZEKE spectrum. When this transition is forbidden or inaccessible, then only the 'back door' mechanism is possible, leading to apparent violations of the Franck–Condon principle and hence strong intensities in the forbidden region. This is an important new channel for observing forbidden transitions by way of ZEKE spectroscopy.

Among van der Waals complexes, we have measured benzene–argon and benzene–krypton complexes, as well as complexes of water with phenol, methanol and ethanol. We have also examined the dimer of phenol and mixed phenol–water–argon complexes. We investigated the metal complexes listed below. The nitric oxide dimer has been investigated and we have measured anions in a series of unusual structures such as hydroxyl anions, iron oxide, iron dicarbide and iron carbohydrates. This opens up the possibility of measuring a whole series of interesting species (Table 27.2), particularly as short-lived chemical intermediates.

The measurement technique is now becoming grossly simplified. The surprise is that ionizing a species and analysing the results with some delay reveals a fine structure buried within the current versus wavelength spectrum and that leads to a new direct spectroscopy.

Table 27.2. *Molecules investigated by ZEKE spectroscopy*

	Name	Formula
Small molecules		
	Nitric oxide	NO
	Nitric dioxide	NO_2
	Ammonia	NH_3
	Nitrogen	N_2
	Nitrous oxide	N_2O
	Hydroxyl radical	OH
	Oygen	O_2
	Iodine	I_2
	Hydrogen fluoride	HF
	Hydrogen chloride	HCl
	Hydrogen bromide	HBr
	Hydrogen iodide	HI
	Hydrogen cyanide	HCN
	Monohydrogen sulphide	HS
	Hydrogen sulphide	H_2S
	Water	H_2O
	Carbon monoxide	CO
	Carbon dioxide	CO_2
	Carbon disulphide	CS_2
	Methyl iodide	CH_3I
	Methyl sulphide	CH_3S
	Acetylene	C_2H_2
	Ethyl iodide	C_2H_5
	Triniobium monoxide	Nb_3O
	Vanadium oxide	VO
Aromatic molecules		
	Benzene	C_6H_6
	Pyrazine	$C_4H_4N_2$
	2–Hydroxypyridine	C_5H_6NO
	p-Difluorobenzene	$C_6H_4F_2$
	Chlorobenzene	C_6H_5Cl
	m-Chlorophenol	C_6H_5ClO
	Phenol	C_6H_6O
	Aniline	C_6H_7N
	Phenylsilane	C_6H_8Si
	Diazabicyclooctane (DABCO)	$C_6H_{12}N_2$
	Benzonitrile	C_7H_5N
	2,6–Difluorotoluene	$C_7H_6F_2$
	2–Fluoro-6–chlorotoluene	C_7H_6FCl
	Tropolone	$C_7H_6O_2$
	o-, *m*- and *p*-chlorotoluene	C_7H_7F
	Toluene	C_7H_8
	Phenylacetylene	C_8H_6
	Ethenylbenzene (styrene)	C_8H_8
	p-Dimethoxybenzene	$C_8H_{10}O_2$
	n-Propylbenzene	C_9H_{12}
	p-n-Propylaniline	$C_9H_{13}N$
	Naphthalene	$C_{10}H_8$
	Naphthol	$C_{10}H_8O$
	Dibenzodioxine	$C_{12}H_8O_2$

Table 27.2. (*cont.*)

Name	Formula
9–Hydroxyphenalenone	$C_{13}H_8O_2$
Tolane	$C_{14}H_{10}$
Anthracene	$C_{14}H_{10}$
trans-Diphenylethene (stilbene)	$C_{14}H_{12}$
9,10–Dihydrophenanthrene	$C_{14}H_{12}$
Perylene	$C_{20}H_{12}$

Van der Waals complexes

Benzene–argon	C_6H_6Ar
Benzene–krypton	C_6H_6Kr
p-Difluorobenzene–argon	$C_6H_4F_2Ar$
Aniline–argon	C_6H_7NAr
Phenylsilane–argon	C_6H_8SiAr
Benzonitrile–argon	C_7H_5NAr
Toluene–argon	C_7H_8Ar
Phenylacetylene–argon	C_8H_6Ar
Ethenylbenzene–argon	C_8H_8Ar
p-Dimethoxybenzene–argon	$C_8H_{10}O_2Ar$
Naphthalene–argon	$C_{10}H_8Ar$
Dibenzodioxine–argon	$C_{12}H_8O_2Ar$
Dibenzodioxine–krypton	$C_{12}H_8O_2Kr$
Anthracene–argon	$C_{14}H_{10}Ar$
Iodine–argon	I_2Ar
Nitric oxide–argon	$NOAr$
Aluminium–argon	$AlAr$

H-bonded complexes

Phenol–water	$C_6H_8O_2$
Phenol–methanol	$C_7H_{10}O_2$
Aniline–methane	$C_7H_{11}N$
Phenol–ethanol	$C_8H_{12}O_2$
Phenol–dimethylether	$C_8H_{12}O_2$
Phenol dimer	$C_{12}H_{10}O_2$

Mixed complexes

Phenol–water–argon	$C_6H_8O_2Ar$

Metal complexes

Silver dimer	Ag_2
Aluminium dimer	Al_2
Vanadium dimer, trimer and tetramer	V_2, V_3, V_4

Molecular complexes

Nitric oxide dimer	N_2O_2

Radicals

Methyl	CH_3
Benzyl	C_7H_7

Anions/neutral species

Small molecule	Hydroxyl radical	OH
	Monohydrogen sulphide	HS
	Iron oxide	FeO
	Iron dicarbide	FeC_2

Table 27.2. (*cont.*)

	Name	Formula
	Hydrogen diiodide	IHI
	Hydrogen dibromide	BrHBr
Van der Waals complexes	Argon–bromine	ArBr
	Argon–iodine	ArI
	Krypton–iodine	KrI
	Carbon dioxide–iodine	CO_2I
	Methyl iodide–iodine	CH_3I I
Clusters and complexes	Argon clusters–bromine	Ar_nBr
	Argon clusters–iodine	Ar_nI
	Silver trimer	Ag_3
	Gold dimer	Au_2
	Silicon dimer, trimer and tetramer	Si_2, Si_3, Si_4
	Germanium dimer	Ge_2
	Carbon clusters	C_5, C_6

28

Epilogue

Clearly this book represents only a momentary glimpse into the development of a new spectroscopy based on a new effect, ZEKE. Though the unusual character of ZEKE states is still being studied, there can be little question of their existence, character and usefulness. There is already a rich body of literature with highly useful ZEKE spectra. In addition and quite apart form the practical usefulness of the method the physics of the ZEKE process involves so much unexpected and new thinking that it has aroused fascination in its own right and, no doubt, prescriptions for a yet wider range of experiments.

ZEKE spectroscopy has been good for many surprises and indeed turns out to be quite counterintuitive (Table 28.1) in ways that the foregoing pages attempted to explain. The field is very young, so there are doubtless many more surprises yet to come.

ZEKE spectroscopy will carve out a niche that will no doubt be complementary to and quite different from the traditional spectroscopies. Not only does it, for the first time, provide laser resolution in ion spectroscopy but it is even showing different excitation paths from photoelectron spectroscopy. ZEKE experiments not only encompass a 1000-fold improved resolution and high accuracy, but also, through the exploitation of new aspects of channel coupling that are unique to this spectroscopy, allow one to observe a richness of quasi-forbidden transitions for standard systems, with the added feature of coincidence of mass spectrometry, to look at ZEKE spectra of fragile complexes and even to observe the ephemeral transition state of chemical reactions, yielding a direct spectroscopic technique. These fragile species, which are of essential importance in the understanding of chemical processes, are seen directly in ZEKE. Perhaps the greatest growth will be in the ground-state spectroscopy of transitory short-lived intermediates of chemical reactions, free radicals and the like. Much

Table 28.1. *Counterintuitive aspects of ZEKE spectroscopy*

---

50 µs ZEKE states are up to 19 eV above the ground state buried in the ionization continuum

Photodecomposition of intact ZEKE states occurs without electron ejection

All transitions are possible; there are no selection-rule restrictions

Extremely strong coupling of all stacks of Rydberg states is dominant

Violation of the Franck–Condon principle occurs

Red-degradation of spectra occurs

Violation of the Born–Oppenheimer picture occurs: the inverted BO is a better description for the ZEKE region

---

remains unknown about the excited states of molecular ions and their complexes.

Clearly one of the great new interesting directions to expect in ZEKE spectroscopy is the addition also of larger timescales to ZEKE spectroscopy. I think this is the next leap forwards, looking at ZEKE in time directly, in a pump–probe short-time coincidence experiment, or in an ICR in a very long experiment looking at soft coupling and long lifetimes. Protein dynamics, for example, teaches us that one must look at all ranges of time scales[239–243].

Study of the liquid and solid states with ZEKE spectroscopy also appears within reach by employing modern tunable picosecond and femtosecond lasers. Femtochemistry joined to ZEKE spectroscopy is starting to produce results. In a femtosecond experiment, all vibrations are frozen in time, so that a transition occurs from frozen nuclei. If the electron is now analysed prior to phonon relaxation, there is a real chance that one might see the chemical signature of the species to hand[244,245].

Clearly, femtosecond, frozen-core experiments are interesting because you then are exciting at 0 K and therefore observing the motion starting from a frozen state. These states, upon excitation, are not even classically vibrating so no excitation is involved. One can then see competitions to these first vibrations.

ZEKE techniques might be of particular interest for surface states in which such phonon uncoupling would be of interest. It must be remembered that also surfaces can produce image states, which is the analogy to the molecular Rydberg state. Similarly, it is expected that there will be discrete states below the vacuum level with strongly enhanced lifetimes.

State-selective experiments also can now be carried out with laser accuracy in coincidence experiments. Now that kilohertz lasers are becoming available, the entire apparatus of coincidence methods will become available for ZEKE spectroscopy. In this way one can study process involving decomposition of highly state-selected ions and neutral species or undertake precise state-selected ion–molecule bimolecular reactions.

No doubt, history has shown that the applications will always be more numerous and more interesting than predicted. The short history of ZEKE spectroscopy up to now has borne this out.

We have looked at the interaction between the intramolecular and the van der Waals modes, which is of great interest, particularly now that we can assign all these new modes, be they forbidden or not. Vibrational predissociation, isomerization and the various types of chemical reactions that can take place here are certainly of considerable interest. The same statement can be extended to the case of clusters, their bonding and reactivity. We can investigate clusters of atoms, for example of silver. I have discussed the work concerning iodine solvated with CO_2, which is very interesting. Even apparently non-specific noble gas atoms attached to molecules exhibit a strong directionality and specific length in their van der Waals bonds. Clearly, interaction forces of van der Waals type on argon are not just characterized by traditional 'van der Waals radii' and we are learning from *ab initio* calculations that simple spherical molecular modelling can produce gross errors. We mentioned hydrogen-bonded systems as an example of soft-bond ZEKE spectroscopy. We go from binding energies of the old world of strong bonds now to the new world of soft bonds with the partial reversibility required by biological systems. We can think about complex biological systems such as gramicidin (Fig. 28.1) and biomolecular clusters of nucleic acid fragments and about solvating these clusters with individual or many solvent molecules. The question of charge in biological structures appears to attain pivotal importance. This certainly is all an area in which interesting insights are expected from ZEKE spectroscopy. Perhaps, to guess at future directions, one might think about experiments with these soft bonds, forbidden transitions, reactive intermediates, extremely short-lived transition states, free radicals, surfaces and much more (Fig. 28.2).

We have talked about pulsed-field ionization of ZEKE states as the way of looking at these states just hugging each transition from below. This has brought an immense improvement in resolution, an accuracy only limited by the laser employed, and observation of many new cross-sections due to channel couplings that we had not seen before. It would be fascinating to extend this work by going to delayed fields, which are particularly

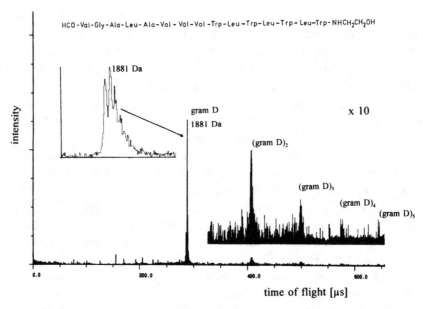

Fig. 28.1 Gramicidin D (gram D). This demonstrates the facile vaporization and laser ionization of this polypeptide, showing the range of molecular system amenable to study.

interesting when we do the new high-energy synchrotron experiments. The new high-resolution synchrotron work at the many new storage rings in Berlin, Berkeley, etc. provides a whole new field of opportunities. ZEKE resolution does not depend on what initial energies you have, but rather it is simply limited by the band width of the light sources you have and nothing else. The instrumental resolution does not degrade with energy, but remains the same. This should be interesting also for liquid jets, which are now being utilized in a number of laboratories. Second-harmonic generation, as well as sum-frequency generation, as in the work of Shen[246] at Berkeley, would be interesting to combine with ZEKE techniques. It would be very interesting to compare HREELS experiments and other, similar surface experiments with ZEKE results, particularly for clusters. Our work seems to show now that the cluster experiments give results very analogous to those from HREELS experiments. This is of interest in view of size definitions of cluster systems. This would be an interesting correspondence of clusters with surfaces. HREELS-type experiments can be conducted easily for more complex systems, e.g. Fe, Fe_2, Fe_3 and FeC_n. Free-radical ions have been examined before by many workers such as Oka[247] at Chicago and molecular ions by Saykally and co-workers[248,249] at Berkeley.

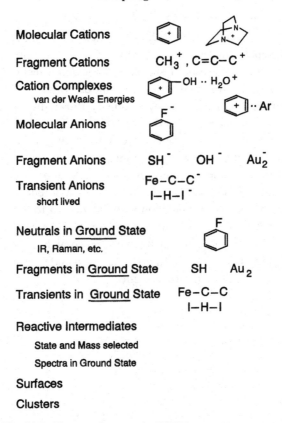

Fig. 28.2 Classes of system in ZEKE spectroscopy.

ZEKE, however, might constitute an interesting method for larger systems that might be more difficult to analyse from a mixture. This new vista certainly would include polyatomic ions, which can be measured in ZEKE experiments together with mass selection.

Since the spectral resolution of ZEKE spectroscopy depends only on the availability of high laser resolution, it is sensitive to the whole new technology of lasers that is emerging. The whole new world of the optical parametric oscillator (OPO) technology which has just developed during the last few years into a routine scannable tool is just one candidate. Another is the whole world of measuring the far infra-red spectrum by scanning – up to some 10 μm – and soon to 20 μm with silver germanium sulphide and selenide. This will be a unique capability for the entire field of anion spectroscopy. Dissociation spectroscopy will become very, very interesting. Two-colour scannable laser systems are now becoming commercially

available. This in turn will make new experiments accessible to the picosec-
ond- and femtosecond-tunable technology in a pump–probe configuration.
When high-repetition-rate lasers become available, we can extend ZEKE
spectroscopy into the realm of standard coincidence technology and thus
look both at the electron and at the ion simultaneously. Such simultaneous
methods can be performed at a lower density, thus reducing the loading of
the system, which is a particularly important issue for biological systems.
Under these conditions we are free from all these non-linear effects that
often so strongly interfere with clean measurements that they make some
measurements in biochemistry impossible. Furthermore, this would permit
us to conduct ZEKE coincidence experiments with energy selection, i.e.
whereas before threshold ionization experiments led to PIPECO, this can
now be done with ZEKE accuracy, precision and the new selection rules.
As I have tried to indicate, ZEKE spectroscopy has expanded into a
number of different laboratories across the world and the spread of the field
is a testimony to the breadth of this effect.

List of acronyms

EELS	ELectron energy-loss spectroscopy
HREELS	High-resolution electron energy-loss spectroscopy.
IBO	Inverse Born–Oppenheimer
MATI	Mass-analysed threshold ionization. The same process as the ZEKE method except with the emphasis that ions rather than electrons are detected.
PDS	Photodetachment spectroscopy. This is a scan, just as in PIE, of the electron current as a function of photon energy, but here for detachment of the electron from an anion.
PES	Photoelectron spectroscopy. Excitation with a fixed frequency above the ionization energy and velocity analysis of the thus-emitted electrons.
PFI	Pulsed field ionization.
PIE	Photo-ionization efficiency. A plot of the total photoelectron current against the excess photon energy above the ionization energy
PIPECO	Photo-ion-photoelectron coincidence.
REMPI	Resonant multiphoton ionization. A technique, usually of two-photon spectroscopy, in which the first photon goes to provide a resonant intermediate state and the second brings about ionization.
RETOF	Reflecting mode of a time-of-flight mass spectrometer for increased resolution and for the detection of metastable species kinetics.
TOF-MS	Time-of-flight mass spectroscopy.
TPES	Threshold photoelectron spectroscopy. Measurement of thresholds for the production of vibronic states of the ion as detected by steradiancy analysis. Steradiancy analysis detects

the lack of steric divergence of electrons emitted at threshold. As a result such threshold electrons can be successfully transmitted through a long pipe.

ZEKE   **Ze**ro **k**inetic **e**nergy spectroscopy of photoelectrons emitted by delayed fields from the highest range of Rydberg states. The spectrum thus obtained reflects the vibronic and rotational states of the cation directly. In anion ZEKE spectroscopy the delayed ionization probes the neutral ground state.

ZEKE bibliography

A list of references on ZEKE spectroscopy had originally been compiled by Dr
T. G. Wright of the University of Southampton and is being continued by Dr
W. J. Knott in Munich. It is kept up to date by entries and submissions via the
World Wide Web.
http://www.chemie.tu-muenchen.de/zeke
or
http://zeke.chemie.tu-muenchen.de

Appendix

High-precision ionization potentials obtained with ZEKE spectroscopy

A.1 Cations

	Molecule	IP (cm^{-1})	Precision (cm^{-1})	Reference
Ag$_2$	silver dimer	61747	3	205
Al·Ar	aluminium–argon	47423	3	250
Al$_2$	aluminium dimer	48304	16	251
CO	carbon monoxide	113025.6	1.5	252
CO$_2$	carbon dioxide	111111	3	253
CS$_2$	carbon disulfide	81286	5	254
CH$_2$O	formaldehyde	87792	3	1
CH$_3$	methyl	79349	3	82
CH$_3$I	methyl iodide	76934	5	255, 256
CD$_3$I	methyl iodide-d$_3$	76958	5	256
CH$_3$S	methyl sulphide	74726	8	257
C$_2$H$_2$	acetylene	91952	2	258
C$_2$H$_5$I	ethyl iodide	75406	5	259
C$_4$H$_4$N$_2$	pyrazine	74903	–	260
C$_5$H$_6$NO	2-pyridinone	68137	5	261
C$_5$H$_6$NO	2-hydroxypyridinol	72093	5	261
C$_6$H$_5$Cl	chlorobenzene	73170	5	262
C$_6$H$_5$Cl·Ar	chlorobenzene–argon	72984	5	263
C$_6$H$_4$F$_2$	p-difluorobenzene	73872	3	77, 264
C$_6$H$_4$F$_2$·Ar	p-difluorobenzene–argon	73637	3	265
C$_6$H$_5$F	fluorobenzene	74238	4	266
C$_6$H$_5$F·Ar	fluorobenzene–argon	74011	4	266
C$_6$H$_5$F·Ar$_2$	fluorobenzene–argon$_2$	73816	4	266
C$_6$H$_6$	benzene	74556.1	0.3	46
		74556.575	0.05	267
C$_6$D$_6$	benzene-d$_6$	74583.507	0.05	267
C$_6$H$_6$·Ar	benzene–argon	74383	2	48, 268
C$_6$H$_6$·Ar$_2$	benzene–argon$_2$	74221	2	269
C$_6$H$_6$Kr	benzene–krypton	74321	2	123
C$_6$H$_6$O	phenol	68625	4	270

A.1 Cations (*cont.*)

Molecule		IP (cm^{-1})	Precision (cm^{-1})	Reference
C_6H_5DO	phenol-d$_1$	68610	4	270
$C_6H_6O\cdot H_2O$	phenol–water	64024	4	270
$C_6H_5DO\cdot H_2O$	phenol-d$_1$–water	64066	4	270
$C_6H_6O\cdot HDO$	phenol–water-d$_1$	63997	4	270
$C_6H_5DO\cdot HDO$	phenol-d$_1$–water-d$_1$	64038	4	270
$C_6H_6O\cdot D_2O$	phenol–water-d$_2$	63973	4	270
$C_6H_5DO\cdot D_2O$	phenol-d$_1$–water-d$_2$	64017	4	270
$C_6H_6O\cdot H_2O\cdot Ar$	phenol–water–argon	63899	4	271
$C_6H_6O\cdot CH_4O$	phenol–methanol	63207	4	272
$C_6H_6O\cdot C_2H_6O$	phenol–ethanol	62901	5	273
$C_6H_6O\cdot C_2H_6O$	phenol–dimethylether	62604	5	274
C_6H_5ClO	*cis-m*-chlorophenol	69810	10	275
C_6H_5ClO	*trans-m*-chlorophenol	70027	10	275
C_6H_7N	aniline	62268	4	276
$C_6H_7N\cdot Ar$	aniline–argon	62157	4	276
$C_6H_7N\cdot Ar_2$	aniline–argon$_2$	62049	4	276
$C_6H_7N\cdot CH_4$	aniline–methane	62112	–	277
C_6H_8Si	phenylsilane	73680	5	278
$C_6H_8Si\cdot Ar$	phenylsilane–argon	73517	5	278
$C_6H_8Si\cdot Ar_2$	phenylsilane–argon$_2$	73359	5	278
$C_6H_{12}N_2$	diazabicyclooctane (DABCO)	22254.121	0.023	279
C_7H_5N	benzonitrile	78490	2	280
$C_7H_5N\cdot Ar$	benzonitrile–argon	78241	4	280
$C_7H_5N\cdot Ar_2$	benzonitrile–argon$_2$	78007	4	280
$C_7H_6F_2$	2,6-difluorotoluene	73674	5	281
C_7H_6FCl	2-fluoro-6-chlorotoluene	72943	5	282
$C_7H_6O_2$	tropolone	68365	5	283
C_7H_7	benzyl	58465	5	84
$C_7H_5D_2$	benzyl-d$_2$	58410	5	84
C_7D_7	benzyl-d$_7$	58382	5	84
C_7H_7F	*o*-fluorotoluene	71858	–	284
C_7H_7F	*m*-fluorotoluene	72000	–	284
C_7H_7F	*p*-fluorotoluene	70930	–	284
C_7H_8	toluene	71199	5	285
$C_7H_8\cdot Ar$	toluene–argon	71033	5	285
$C_7H_8\cdot Ar_2$	toluene–argon$_2$	70871	5	285
C_8H_6	phenylacetylene	71175	5	286
$C_8H_6\cdot Ar$	phenylacetylene–argon	71027	5	286
C_8H_7N	indole	62592	4	287
$C_8H_7N\cdot Ar$	indole–argon	62504	6	287
C_8H_8	ethenylbenzene (styrene)	68267	5	286
$C_8H_8\cdot Ar$	ethenylbenzene–argon	68151	5	286
$C_8H_{10}O_2$	*cis-p*-dimethoxybenzene	60774	7	288
$C_8H_{10}O_2$	*trans-p*-dimethoxybenzene	60563	7	288
$C_8H_{10}O_2\cdot Ar$	*cis-p*-dimethoxybenzene–argon	60687	7	288
$C_8H_{10}O_2\cdot Ar$	*trans-p*-dimethoxybenzene–argon	60479	7	288

A.1 Cations (*cont.*)

	Molecule	IP (cm^{-1})	Precision (cm^{-1})	Reference
$C_8H_{10}O_2 \cdot Ar_2$	*cis-p*-dimethoxybenzene–argon$_2$	60509	7	288
$C_8H_{10}O_2 \cdot Ar_2$	*trans-p*-dimethoxybenzene–argon$_2$	60295	7	288
C_9H_{12}	*trans*-n-propylbenzene	70278	8	289
C_9H_{12}	*gauche*-n-propylbenzene	70420	8	289
$C_9H_{13}N$	*anti-para*-n-propylaniline	59717	3	290
$C_9H_{13}N$	*gauche-para*-n-propylaniline	59793	3	290
$C_{10}H_8$	naphthalene	65687	7	291
$C_{10}H_8Ar$	naphthalene–argon	65607	3	292
$C_{10}H_8O$	*cis*-1-naphthol	62918	8	293
$C_{10}H_8O$	*trans*-1-naphthol	62637	8	293
$C_{10}H_8O$	*cis*-2-naphthol	63670	8	293
$C_{10}H_8O$	*trans*-2-naphthol	63189	8	293
$C_{12}H_8O_2$	dibenzo-*p*-dioxine	61283	2	294
$C_{12}H_8O_2 \cdot Ar$	dibenzo-*p*-dioxine–argon	61071	2	294
$C_{12}H_8O_2 \cdot Kr$	dibenzo-*p*-dioxine–krypton	61005	2	294
$C_{12}H_9N$	carbazole	61426	4	295
$C_{12}H_{10}O_2$	phenol dimer	63649	4	296
$C_{14}H_{10}$	tolane	63917	3	297
$C_{13}H_8O_2$	9-hydroxyphenalenone (9-HPO-h)	65338	5	283
$C_{13}H_7DO_2$	9-hydroxyphenalenone (9-HPO-d) (deuterated)	65350	5	283
$C_{14}H_{10}$	anthracene	59872	5	298
$C_{14}H_{10} \cdot Ar$	anthracene–argon (1+0)	59807	5	298
$C_{14}H_{10} \cdot Ar$	anthracene–argon (1+0)	59825	5	298
$C_{14}H_{10} \cdot Ar_2$	anthracene–argon$_2$ (1+1)	59757	5	298
$C_{14}H_{10} \cdot Ar_2$	anthracene–argon$_2$ (2+0)	59774	5	298
$C_{14}H_{10} \cdot Ar_3$	anthracene–argon$_3$ (2+1)	59695	5	298
$C_{14}H_{10} \cdot Ar_4$	anthracene–argon$_4$ (2+2)	59606	5	298
$C_{14}H_{10} \cdot Ar_4$	anthracene–argon$_4$ (3+1)	59660	5	298
$C_{14}H_{10} \cdot Ar_5$	anthracene–argon$_5$ (3+2)	59565	5	298
$C_{14}H_{12}$	*trans*-diphenylethene (stilbene)	61756	–	299
$C_{14}H_{12}$	9,10-dihydrophenanthrene	63645	5	300
HBr	hydrogen bromide	94098.9	1	301
HCN	hydrogen cyanide	109750	2	302
HCl	hydrogen chloride	102802.8	2	303
HF	hydrogen fluoride	129422.4	1	304
HI	hydrogen iodide	–	–	305
HO	hydroxy	104989	2	83
DO	hydroxy-d	105085	2	83
HS	monohydrogen sulphide	84057.5	3	306
H_2O	water	101766	2	307
D_2O	water-d$_2$	101916	2	307
H_2S	hydrogen sulphide	84432	2	308
I_2	iodine	75069	2	309
$I_2 \cdot Ar$	iodine–argon	74523	2	310
NH_3	ammonia	82159	1	55, 137

A.1 Cations (*cont.*)

	Molecule	IP (cm^{-1})	Precision (cm^{-1})	Reference
ND$_4$	ammonium-d$_4$	37490.7	1.5	311
NO	nitric oxide	74719.0	0.5	71
NO·Ar	nitric oxide–argon	73869	6	312
NO$_2$	nitric dioxide	77315.9	1	313
N$_2$	nitrogen	125668	0.25	314
N$_2$O	nitrous oxide	103963	5	315
(NO)$_2$	nitric oxide dimer	70350	5	316
Na$_3$	sodium trimer	31363	5	317
Na·H$_2$O	sodium–water	35323	10	318
Na·D$_2$O	sodium–water-d$_2$	35249	10	318
Na·NH$_3$	sodium–ammonia	34435	10	318
Na·ND$_3$	sodium–ammonia-d$_3$	34368	10	318
Nb$_3$C$_2$	triniobium dicarbide	40639.0	3	319
Nb$_3$N$_2$	triniobium dinitride	43901.7	5	320
Nb$_3$O	triniobium monoxide	44578	3	321
O$_2$	oxygen	97348	2	57
VO	vanadium oxide	58383	5	322
V$_2$	vanadium dimer	51271.14	0.5	323
V$_3$	vanadium trimer	44342	3	324
V$_4$	vanadium tetramer	45664	3	324
Y$_2$	diyttrium	40131	2	325

A.2 Anions

	Molecule	EA (cm^{-1})	Precision (cm^{-1})	Reference
Ag$_3$	silver trimer	–	–	157
Ar·Br	argon–bromine	27429.6	3	167
Ar·Br$_2$	argon–bromine$_2$	27722.4	3	326
Ar·Br$_3$	argon–bromine$_3$	27994.6	3	326
Ar·I	argon–iodine	24888.3	3	167, 326
Ar·I$_2$	argon–iodine$_2$	25100.9	3	326
Ar·I$_3$	argon–iodine$_3$	25303.0	3	326
Ar·I$_4$	argon–iodine$_4$	25502.2	3	326
Au$_2$	gold dimer	15642	4	157
Au$_6$	gold hexamer	16541	17	327
C$_5$	carbon$_5$	23013	10	162
C$_6$	carbon$_6$	33714	8	164
CH$_3$I·I	methyl iodide–iodine	27427	10	328
FeO	iron oxide	12054	5	329
FeC$_2$	iron dicarbide	15950	5	170
Ge$_2$	germanium dimer	16727	10	177

A.2 Anions (*cont.*)

	Molecule	EA (cm^{-1})	Precision (cm^{-1})	Reference
I	iodine atom	24673	3	330
I·CO$_2$	carbon dioxide–iodine	26011	5	182
InP$_2$	indium diphosphide	13042	8	166
In$_2$P	diindium phosphide	19357	8	166
Kr·I	krypton–iodine	25020.6	6	167
OH	hydroxyl radical	14741.02	3	331
SH	monohydrogen sulphide	18688	16	330
Si$_2$	silicon dimer	17760	8	160
Si$_3$	silicon trimer	18564.93	10	161
Si$_4$	silicon tetramer	17502	8	332

References

1. K. Müller-Dethlefs, E. W. Schlag, E. R. Grant, K. Wang, and V. B. McKoy, in *Advances in Chemical Physics*, edited by I. Prigogine and S. A. Rice (Wiley, New York, 1995).
2. R. D. Levine, *Advances in Chemical Physics. Proceedings of the XXth Solvay Conference, Brussels* **in press** (1997).
3. J. Jortner and M. Bixon, *J. Chem. Phys.* **102**, 5636 (1995).
4. S. H. Lin, Y. Fujimura, H. J. Neusser, and E. W. Schlag, *Multiphoton Spectroscopy of Molecules* (Academic Press, New York, 1984).
5. M. Göppert, *Naturwiss.* **17**, 932 (1929).
6. M. Göppert-Mayer, *Ann. Phys.* **9**, 273 (1931).
7. U. Boesl, H. J. Neusser, and E. W. Schlag, *Z. Naturforsch. A* **33**, 1546 (1978).
8. L. Zandee, R. B. Bernstein, and D. A. Lichtin, *J. Chem. Phys.* **69**, 3427 (1978).
9. J. H. Callomon, T. M. Dunn, and I. M. Mills, *Phil. Trans. R. Soc. London Ser. A* **259**, 499 (1966).
10. W. Forst, *Theory of Unimolecular Reactions* (Academic Press, New York and London, 1973).
11. R. A. Marcus, *J. Chem. Phys.* **43**, 2658 (1965).
12. P. Hobza, H. L. Selzle, and E. W. Schlag, *J. Phys. Chem.* **100**, 18790 (1996).
13. P. Hobza, H. L. Selzle, and E. W. Schlag, *J. Chem. Phys.* **98**, 5893 (1990).
14. P. Hobza, H. L. Selzle, and E. W. Schlag, *J. Phys. Chem.* **97**, 3937 (1993).
15. P. Hobza, H. L. Selzle, and E. W. Schlag, *J. Am. Chem. Soc.* **116**, 3500 (1994).
16. O. Krätzschmar, H. L. Selzle, and E. W. Schlag, *J. Phys. Chem.* **98**, 3501 (1994).
17. B. Ernstberger, H. Krause, and H. J. Neusser, *Ber. Bunsenges. Phys. Chem.* **97**, 884 (1993).
18. B. A. Mamyrin and D. V. Shmikk, *Sov. Phys. JETP* **49**, 762 (1979).
19. U. Boesl, R. Weinkauf, and E. W. Schlag, *Int. J. Mass Spectrom. Ion Processes* **112**, 121 (1992).
20. U. Boesl, H. J. Neusser, and E. W. Schlag, *J. Am. Chem. Soc.* **103**, 5058 (1981).
21. V. S. Antonov, I. N. Knyazev, V. S. Letokhov, V. M. Matiuk, V. G. Movshev, and V. K. Potapov, *Opt. Lett.* **2**, 37 (1978).
22. U. Boesl, H. J. Neusser, R. Weinkauf, and E. W. Schlag, *J. Phys. Chem.* **86**, 4857 (1982).

23. H. Kühlewind, H. J. Neusser, and E. W. Schlag, *Int. J. Mass Spectrom. Ion Phys.* **51**, 255 (1983).
24. B. Ernstberger, H. Krause, A. Kiermeier, and H. J. Neusser, *J. Chem. Phys.* **92**, 5285 (1990).
25. R. Weinkauf, K. Walter, C. Weickhard, U. Boesl, and E. W. Schlag, *Z. Naturforsch. A* **44**, 1219 (1989).
26. E. Riedle, Th. Knittel, Th. Weber, and H. J. Neusser, *J. Chem. Phys.* **91**, 4555 (1989).
27. H. J. Neusser and H. Krause, *Int. J. Mass Spectrom. Ion Processes* **131**, 211 (1994).
28. B. Ernstberger, H. Krause, and H. J. Neusser, *Z. Phys. D: Atoms, Molecules, Clusters* **20**, 189 (1991).
29. H. Krause, B. Ernstberger, and H. J. Neusser, *Chem. Phys. Lett.* **184**, 411 (1991).
30. E. Miescher, Y. T. Lee, and P. Gürtler, *J. Chem. Phys.* **68**, 2753 (1978).
31. E. Sekreta, K. S. Viswanathan, and J. P. Reilly, *J. Chem. Phys.* **90**, 5349 (1989).
32. D. Villarejo, R. R. Herm, and M. G. Inghram, *J. Chem. Phys.* **46**, 4995 (1967).
33. P. M. Guyon, T. Baer, L. F. A. Ferreira, I. Nenner, A. Tabche-Fouhaile, R. Botter, and T. R. Govers, *J. Phys. B: Atomic Mol. Phys.* **11**, L141 (1978).
34. W. B. Peatman, *Ph. D. Thesis* (Northwestern University, Evanston, 1969).
35. G. Reiser, W. Habenicht, K. Müller-Dethlefs, and E. W. Schlag, *Chem. Phys. Lett.* **152**, 119 (1988).
36. L. Ya. Baranov, *Ph. D. Thesis* (Hebrew University, Jerusalem, 1996).
37. R. Neuhauser and H. J. Neusser, *Chem. Phys. Lett.* **253**, 151 (1996).
38. T. Baer, *Ann. Rev. Phys. Chem.* **40**, 637 (1989).
39. C.-W. Hsu, K. T. Lu, M. Evans, Y. J. Chen, C. Y. Ng, and P. Heimann, *J. Chem. Phys.* **105**, 3950 (1996).
40. L. Zhu and P. Johnson, *J. Chem. Phys.* **94**, 5769 (1991).
41. Sh.-I. Sato and K. Kimura, *Chem. Phys. Lett.* **249**, 155 (1996).
42. E. Miescher, *Can. J. Phys.* **54**, 2074 (1976).
43. D. T. Biernacki, S. D. Colson, and E. E. Eyler, *J. Chem. Phys.* **89**, 2599 (1988).
44. L. A. Chewter, M. Sander, K. Müller-Dethlefs, and E. W. Schlag, *J. Chem. Phys.* **86**, 4737 (1987).
45. S. G. Grubb, R. L. Whetten, A. C. Albrecht, and E. R. Grant, *Chem. Phys. Lett.* **108**, 420 (1984).
46. I. Fischer, R. Lindner, and K. Müller-Dethlefs, *J. Chem. Soc. Faraday Trans.* **90**, 2425 (1994).
47. M. A. Duncan, T. G. Dietz, and R. E. Smalley, *J. Chem. Phys.* **75**, 2118 (1981).
48. L. A. Chewter, K. Müller-Dethlefs, and E. W. Schlag, *Chem. Phys. Lett.* **135**, 219 (1987).
49. K. H. Fung, W. E. Henke, T. R. Hays, H. L. Selzle, and E. W. Schlag, *J. Phys. Chem.* **85**, 3560 (1981).
50. O. Dopfer, *Dissertation* (Technische Universität München, 1994).
51. M. Fujii, T. Kakinuma, N. Mikami, and M. Ito, *Chem. Phys. Lett.* **127**, 76 (1986).
52. D. Rieger, *Diplomarbeit* (Technische Universität München, 1990).
53. G. Reiser, O. Dopfer, R. Lindner, G. Henri, K. Müller-Dethlefs, E. W. Schlag, and S. D. Colson, *Chem. Phys. Lett.* **181**, 1 (1991).

54. R. J. Lipert and S. D. Colson, *J. Chem. Phys.* **89**, 4579 (1988).
55. W. Habenicht, G. Reiser, and K. Müller-Dethlefs, *J. Chem. Phys.* **95**, 4809 (1991).
56. J. W. Rabalais, L. Karlsson, L. O. Werme, T. Bergmark, and K. Siegbahn, *J. Chem. Phys.* **58**, 3370 (1973).
57. R. G. Tonkyn, J. W. Winniczek, and M. G. White, *Chem. Phys. Lett.* **164**, 137 (1989).
58. K. P. Huber and G. Herzberg, *Molecular Spectra and Molecular Structure. IV. Spectra of Diatomic Molecules* (Van Nostrand Reinhold, New York, 1979).
59. B. Bühler, *Dissertation* (Albert-Ludwigs Universität Freiburg, 1990).
60. C. Bordas, P. Labastie, J. Chevaleyre, and M. Broyer, *Chem. Phys.* **129**, 21 (1989).
61. H. J. Neusser, *private communication*
62. W. A. Chupka, *J. Chem. Phys.* **98**, 4520 (1993).
63. U. Even, R. D. Levine, and R. Bersohn, *J. Phys. Chem.* **98**, 3472 (1994).
64. M. Sander, *Dissertation* (Technische Universität München, 1987).
65. D. W. Turner, C. Baker, A. D. Baker, and C. R. Brundle, *Molecular Photoelectron Spectroscopy* (Wiley, London, 1970).
66. J. Berkowitz, *Photoabsorption, Photoionization, and Photoelectron Spectroscopy* (Academic Press, New York, 1979).
67. Y. Morioka, Y. Lu, T. Matsui, T. Tanaka, H. Yoshii, T. Hayaishi, and R. I. Hall, *J. Chem. Phys.* **104**, 9357 (1996).
68. K. Müller-Dethlefs, M. Sander, and E. W. Schlag, *Z. Naturforsch. A* **39**, 1089 (1984).
69. K. Müller-Dethlefs, M. Sander, and E. W. Schlag, *Chem. Phys. Lett.* **112**, 291 (1984).
70. R. T. Wiedmann and M. G. White, *J. Chem. Phys.* **102**, 5141 (1995).
71. M. Sander, L. A. Chewter, K. Müller-Dethlefs, and E. W. Schlag, *Phys. Rev. A* **36**, 4543 (1987).
72. H. Rudolph, V. McKoy, and S. N. Dixit, *J. Chem. Phys.* **90**, 2570 (1989).
73. K. Wang and V. McKoy, *Ann. Rev. Phys. Chem.* **46**, 275 (1995).
74. S. Fredin, D. Gauyacq, M. Horani, Ch. Jungen, G. Lefevre, and F. Masnou-Seeuws, *Mol. Phys.* **60**, 825 (1987).
75. V. M. Akulin, G. Reiser, and E. W. Schlag, *Chem. Phys. Lett.* **195**, 383 (1992).
76. P. Rosmus, *private communication*
77. D. Rieger, G. Reiser, K. Müller-Dethlefs, and E. W. Schlag, *J. Phys. Chem.* **96**, 12 (1992).
78. P. Botschwina, *Habilitationsschrift* (Universität Kaiserslautern, 1984).
79. H. Krause and H. J. Neusser, *J. Chem. Phys.* **97**, 5923 (1992).
80. R. Lindner, H. Sekiya, B. Beyl, and K. Müller-Dethlefs, *Angew. Chem.* **105**, 631 (1993).
81. S. R. Long, J. T. Meek, and J. P. Reilly, *J. Chem. Phys.* **79**, 3206 (1983).
82. J. A. Blush, P. Chen, R. T. Wiedmann, and M. G. White, *J. Chem. Phys.* **98**, 3557 (1993).
83. R. T. Wiedmann, R. G. Tonkyn, M. G. White, K. Wang, and V. McKoy, *J. Chem. Phys.* **97**, 768 (1992).
84. G. C. Eiden, F. Weinhold, and J. C. Weisshaar, *J. Chem. Phys.* **95**, 8665 (1991).
85. G. C. Eiden, K.-T. Lu, J. Badenhoop, F. Weinhold, and J. C. Weisshaar, *J. Chem. Phys.* **104**, 8886 (1996).

86. G. C. Eiden and J. C. Weisshaar, *J. Chem. Phys.* **104**, 8896 (1996).

87. D. M. Neumark, K. R. Lykke, T. Andersen, and W. C. Lineberger, *J. Chem. Phys.* **83**, 4364 (1985).

88. R. L. Jackson, A. H. Zimmermann, and J. I. Braumann, *J. Chem. Phys.* **71**, 2088 (1979).

89. K. R. Lykke, R. D. Mead, and W. C. Lineberger, *Phys. Rev. Lett.* **52**, 2221 (1984).

90. C. Bäßmann, R. Käsmaier, G. Drechsler, and U. Boesl, in *AIP Conference Proceedings 329: Resonance Ionisation Spectroscopy (Bernkastle-Kues)*, edited by H.-J. Kluge, J. E. Parks, and K. Wendt (American Institute of Physics, New York, 1995).

91. E. Rabani, R. D. Levine, and U. Even, *J. Phys. Chem.* **98**, 8834 (1994).

92. D. W. Turner and M. I. Al Jobory, *J. Chem. Phys.* **37**, 3007 (1962).

93. M. I. Al-Joboury and D. W. Turner, *J. Chem. Soc.*, 5141 (1963).

94. F. I. Vilesov, B. L. Kurbatov, and A. N. Terenin, *Sov. Phys. Doklady* **6**, 490 (1961).

95. A. Terenin and F. Vilesov, *Adv. Photochem.* **2**, 385 (1964).

96. K. Siegbahn, C. Nordling, A. Fahlman, R. Nordberg, K. Hamrin, J. Hedman, G. Johansson, T. Bergmark, S. Karlsson, I. Lindgren, and B. Lindberg, *ESCA Atomic Molecular and Solid State Structure by Means of Electron Spectroscopy* (Almqvist & Wiksells Boktryckeri AB, Uppsala, 1967).

97. K. Watanabe, *J. Chem. Phys.* **22**, 1564 (1954).

98. D. J. Leahy, K. L. Reid, and R. N. Zare, *J. Chem. Phys.* **95**, 1757 (1991).

99. K. Kimura, S. Katsumata, Y. Achiba, T. Yamazaki and S. Iwata, *Handbook of He I Photoelectron Spectra of Fundamental Organic Molecules* (Halsted Press, New York, 1981).

100. K.-M. Weitzel, J. Mähnert, and M. Penno, *Chem. Phys.* **187**, 117 (1994).

101. K.-M. Weitzel, J. Mähnert, and M. Penno, *Chem. Phys. Lett.* **224**, 371 (1994).

102. K.-M. Weitzel and F. Güthe, *Chem. Phys. Lett.* **251**, 295 (1996).

103. T. Baer, P. M. Guyon, I. Nenner, A. Tabche-Fouhaile, R. Botter, and T. R. Govers, *J. Chem. Phys.* **70**, 1585 (1979).

104. P. Morin, I. Nenner, P. M. Guyon, O. Dutuit, and K. Ito, *J. de Chimie Physique* **77**, 605 (1980).

105. R. N. Zare, *Angular Momentum* (John Wiley & Sons, New York, 1987).

106. H.-J. Dietrich, *Dissertation* (Technische Universität München, 1996).

107. R. Lindner, *Dissertation* (Technische Universität München, 1996).

108. R. Lindner, H.-J. Dietrich, and K. Müller-Dethlefs, *Chem. Phys. Lett.* **228**, 417 (1994).

109. W. A. Chupka, *J. Chem. Phys.* **99**, 5800 (1993).

110. U. Even, M. Ben-Nun, and R. D. Levine, *Chem. Phys. Lett.* **210**, 416 (1993).

111. T. F. Gallagher, *Rep. Prog. Phys.* **51**, 143 (1988).

112. C. R. Mahon, G. R. Janik, and T. F. Gallagher, *Phys. Rev. A* **41**, 3746 (1990).

113. C. Bordas and H. Helm, *Phys. Rev. A* **47**, 1209 (1993).

114. C. Bordas, P. Brevet, J. Chevaleyre, and P. Labastie, *Europhys. Lett.* **3**, 789 (1987).

115. D. Bahatt, U. Even, and R. D. Levine, *J. Chem. Phys.* **98**, 1744 (1993).

116. F. Remacle and R. D. Levine, *J. Chem. Phys.* **105**, 4649 (1996).

117. F. Remacle and R. D. Levine, *J. Chem. Phys.* **104**, 1399 (1996).

118. M. Bixon and J. Jortner, *J. Phys. Chem.* **99**, 7466 (1995).

119. F. Merkt and R. N. Zare, *J. Chem. Phys.* **101**, 3495 (1994).
120. F. Merkt, H. H. Fielding, and T. P. Softley, *Chem. Phys. Lett.* **202**, 153 (1993).
121. F. Merkt and T. P. Softley, *Int. Rev. Phys. Chem.* **12**, 205 (1993).
122. C. Alt, W. G. Scherzer, H. L. Selzle, and E. W. Schlag, *Chem. Phys. Lett.* **224**, 366 (1994).
123. H. Krause and H. J. Neusser, *J. Chem. Phys.* **99**, 6278 (1993).
124. W. G. Scherzer, H. L. Selzle, E. W. Schlag, and R. D. Levine, *Phys. Rev. Lett.* **72**, 1435 (1994).
125. W. G. Scherzer, H. L. Selzle, and E. W. Schlag, *Z. Naturforsch. A* **48**, 1256 (1993).
126. C. Bordas, P. F. Brevet, M. Broyer, J. Chevaleyre, P. Labastie, and J. P. Perrot, *Phys. Rev. Lett.* **60**, 917 (1988).
127. A. Held, L. Ya. Baranov, H. L. Selzle, and E. W. Schlag, *Z. Naturforsch. A* **51**, 1236 (1996).
128. M. J. J. Vrakking and Y. T. Lee, *Phys. Rev. A* **51**, R894 (1995).
129. M. J. J. Vrakking, I. Fischer, D. M. Villeneuve, and A. Stolow, *J. Chem. Phys.* **103**, 4538 (1995).
130. C. Alt, W. G. Scherzer, H. L. Selzle, and E. W. Schlag, *Chem. Phys. Lett.* **240**, 457 (1995).
131. A. Held, L. Ya. Baranov, H. L. Selzle, and E. W. Schlag, *Chem. Phys. Lett.* **267**, 318 (1997).
132. G. I. Nemeth, H. L. Selzle, and E. W. Schlag, *Chem. Phys. Lett.* **215**, 151 (1993).
133. L. Asbrink, E. Lindholm, and O. Edqvist, *Chem. Phys. Lett.* **5**, 609 (1970).
134. J. Neukammer, H. Rinneberg, K. Vietzke, A. König, H. Hieronymus, M. Kohl, and H.-J. Grabka, *Phys. Rev. Lett.* **59**, 2947 (1987).
135. Th. F. Gallagher, *Rydberg Atoms* (Cambridge University Press, Cambridge, 1994).
136. J. W. Hepburn, in *Laser Techniques in Chemistry*, edited by A. Myers and T. R. Rizzo (John Wiley & Sons, New York, 1995).
137. G. Reiser, W. Habenicht, and K. Müller-Dethlefs, *J. Chem. Phys.* **98**, 8462 (1993).
138. Th. Weber, A. von Bergen, E. Riedle, and H. J. Neusser, *J. Chem. Phys.* **92**, 90 (1990).
139. C. E. Alt, W. G. Scherzer, H. L. Selzle, and E. W. Schlag, *Ber. Bunsenges. Phys. Chem.* **99**, 332 (1995).
140. H. Krause and H. J. Neusser, *J. Photochem. Photobiol. A: Chem.* **80**, 73 (1994).
141. F. Merkt, *Discussion Meeting of the Royal Society on Molecular Rydberg Dynamics* (The Royal Society, London, 1996).
142. R. Lindner, H. Sekiya, B. Beyl, and K. Müller-Dethlefs, *Angew. Chem. Int. Ed. Engl.* **32**, 603 (1993).
143. H.-J. Dietrich, R. Lindner, and K. Müller-Dethlefs, *J. Chem. Phys.* **101**, 3399 (1994).
144. A. Held, H. L. Selzle, and E. W. Schlag, *J. Phys. Chem. A* **101**, 533 (1997).
145. A. Mühlpfordt, U. Even, E. Rabani, and R. D. Levine, *Phys. Rev. A* **51**, 3922 (1995).
146. H. Dickinson, D. Rolland, and T. P. Softley, *Trans. R. Soc. London Ser. A* **submitted** (1996).
147. I. M. Waller, T. N. Kitsopoulos, and D. M. Neumark, *J. Phys. Chem.* **94**, 2240 (1990).
148. X. Zhang and P. Chen, *private communication*.

264   *References*

149. E. W. Schlag, B. S. Rabinovitch, and F. W. Schneider, *J. Chem. Phys.* **32**, 1599 (1960).
150. F. Merkt, *Discussion Meeting of the Royal Society on Molecular Rydberg Dynamics* (The Royal Society, London, 1996).
151. D. M. Neumark, *Ann. Rev. Phys. Chem.* **43**, 153 (1992).
152. E. Fermi and E. Teller, *Phys. Rev* **72**, 399 (1947).
153. E. A. Brinkman, S. Berger, J. Marks, and J. I. Braumann, *J. Chem. Phys.* **99**, 7586 (1993).
154. A. S. Mullin, K. K. Murray, C. P. Schulz, and W. C. Lineberger, *J. Chem. Phys.* **97**, 10281 (1993).
155. D. Klar, M.-W. Ruf, and H. Hotop, *Chem. Phys. Lett.* **189**, 448 (1994).
156. J. Randell, S. L. Lunt, G. Mrotzek, D. Field, and J. P. Ziesel, *Chem. Phys. Lett.* **252**, 253 (1996).
157. G. F. Gantefӧr, D. M. Cox, and A. Kaldor, *J. Chem. Phys.* **93**, 8395 (1990).
158. G. R. Burton, C. Xu, C. C. Arnold, and D. M. Neumark, *J. Chem. Phys.* **109**, 2757 (1996).
159. T. N. Kitsopoulos, C. J. Chick, Y. Zhao, and D. M. Neumark, *J. Chem. Phys.* **95**, 1441 (1991).
160. C. C. Arnold, T. N. Kitsopoulos, and D. M. Neumark, *J. Chem. Phys.* **99**, 766 (1993).
161. C. C. Arnold and D. M. Neumark, *J. Chem. Phys.* **100**, 1797 (1994).
162. T. N. Kitsopoulos, C. J. Chick, Y. Zhao, and D. M. Neumark, *J. Chem. Phys.* **95**, 5479 (1991).
163. D. W. Arnold, S. E. Bradforth, T. N. Kitsopoulos, and D. M. Neumark, *J. Chem. Phys.* **95**, 8753 (1991).
164. C. C. Arnold, Y. Zhao, T. N. Kitsopoulos, and D. M. Neumark, *J. Chem. Phys.* **97**, 6121 (1992).
165. C. Xu, E. de Beer, D. W. Arnold, C. C. Arnold, and D. M. Neumark, *J. Chem. Phys.* **101**, 5406 (1994).
166. C. C. Arnold and D. M. Neumark, *Can. J. Phys.* **72**, 1322 (1994).
167. Y. Zhao, I. Yourshaw, G. Reiser, C. C. Arnold, and D. M. Neumark, *J. Chem. Phys.* **101**, 6538 (1994).
168. D. M. Neumark, *Acc. Chem. Res.* **26**, 33 (1993).
169. C. Bäßmann, *Dissertation* (Technische Universität München, 1996).
170. G. Drechsler, C. Bäßmann, U. Boesl, and E. W. Schlag, *Z. Naturforsch. A* **49**, 1256 (1994).
171. G. Drechsler, U. Boesl, C. Bäßmann, and E. W. Schlag, *J. Chem. Phys.* **in press** (1997).
172. G. V. Hartland and H.-L. Dai, in *Molecular Dynamics and Spectroscopy by Stimulated Emission Pumping,* edited by H.L. Dai and R. W. Field (World Scientific, Singapore, 1995).
173. E. P. Wigner, *Phys. Rev.* **73**, 1002 (1948).
174. H. Hotop, T. A. Patterson, and W. C. Lineberger, *J. Chem. Phys.* **60**, 1806 (1974).
175. D. M. Neumark, *Acc. Chem. Res.* **26**, 34 (1993).
176. P. A. Schulz, R. D. Mead, P. L. Jones, and W. C. Lineberger, *J. Chem. Phys.* **77**, 1153 (1982).
177. C. C. Arnold, C. Xu, G. R. Burton, and D. M. Neumark, *J. Chem. Phys.* **102**, 6982 (1995).
178. G. Drechsler, C. Bäßmann, U. Boesl, and E. W. Schlag, *J. Mol. Struct.* **348**, 337 (1995).

179. S. D. Moustaizis, M. Tatarakis, C. Kalpouzos, and C. Fotakis, *Appl. Phys. Lett.* **60**, 1939 (1992).
180. G. Drechsler, *Dissertation* (Technische Universität München, 1995).
181. a) G. Markovich, S. Pollak, R. Giniger, and O. Cheshnovsky, in *Reaction Dynamics in Clusters and Condensed Phases*, edited by J. Jortner *et al.* (Kluwer Academic Publishers, 1994). b) D. W. Arnold, S. E. Bradforth, E. H. Kim, and D. M. Neumark, *J. Chem. Phys.* **102**, 3510 (1995). c) L.-S. Wang, J. Conceicao, C. Jin, and R. E. Smalley, *Chem. Phys. Lett.* **182**, 5 (1991).
182. Y. Zhao, C. C. Arnold, and D. M. Neumark, *J. Chem. Soc. Faraday Trans.* **89**, 1449 (1993).
183. C. H. Becker, P. Casavecchia, and Y. T. Lee, *J. Chem. Phys.* **69**, 2377 (1978).
184. C. H. Becker, P. Casavecchia, and Y. T. Lee, *J. Chem. Phys.* **70**, 2986 (1979).
185. C. H. Becker, J. J. Valentini, P. Casavecchia, S. J. Sibener, and Y. T. Lee, *Chem. Phys. Lett.* **61**, 15 (1979).
186. P. Casavecchia, G. He, R. K. Sparks, and Y. T. Lee, *J. Chem. Phys.* **75**, 710 (1981).
187. P. Casavecchia, G. He, R. K. Sparks, and Y. T. Lee, *J. Chem. Phys.* **77**, 1878 (1982).
188. M. F. Golde and B. Thrush, *J. Chem. Phys.* **29**, 486 (1974).
189. T. Baumert, C. Röttgermann, C. Rothenfusser, R. Thalweiser, V. Weiss, and G. Gerber, *Phys. Rev. Lett.* **69**, 1512 (1992).
190. A. H. Zewail, *Photochemistry and Photobiology, Volume 2* (Harwood Academic Publishers, New York, 1983).
191. J. Manz and L. Wöste, *Femtosecond Chemistry* (VCH Weinheim, New York, 1995).
192. B. S. Rabinovitch, R. F. Kubin, and R. E. Harrington, *J. Chem. Phys.* **38**, 405 (1963).
193. J. D. Rynbrandt and B. S. Rabinovitch, *J. Phys. Chem.* **75**, 2164 (1971).
194. R. Weinkauf, P. Aicher, G. Wesley, J. Grotemeyer, and E. W. Schlag, *J. Phys. Chem.* **98**, 8381 (1994).
195. S. Glasstone, K. J. Laidler, and H. Eyring, *The Theory of Rate Processes* (McGraw-Hill Book Company, New York, 1941).
196. I. Fischer, M. J. J. Vrakking, D. M. Villeneuve, and A. Stolow, *Chem. Phys.* **207**, 331 (1996).
197. R. S. Berry and S. E. Nielsen, *J. Chem. Phys.* **49**, 116 (1968).
198. G. P. Bryant, Y. Jiang, M. Martin, and E. R. Grant, *J. Phys. Chem.* **96**, 6875 (1992).
199. G. Gerber and B. Bühler, *unpublished results*
200. Ch. Yeretzian, R. H. Hermann, H. Ungar, H. L. Selzle, E. W. Schlag, and S. H. Lin, *Chem. Phys. Lett.* **239**, 61 (1995).
201. J. W. Hepburn, *Discussion Meeting of the Royal Society on Molecular Rydberg Dynamics* (The Royal Society, London, 1996).
202. Ch. Jungen, *Molecular Applications of Quantum Defect Theory* (Institute of Physics Publishing, Bristol, 1996).
203. R. S. Berry, *J. Chem. Phys.* **45**, 1228 (1966).
204. W. B. Peatman, *J. Chem. Phys.* **64**, 4368 (1976).
205. G. I. Nemeth, H. Ungar, C. Yeretzian, H. L. Selzle, and E. W. Schlag, *Chem. Phys. Lett.* **228**, 1 (1994).
206. U. Fano, *J. Opt. Soc. Am.* **65**, 979 (1975).
207. M. Braunstein, V. McKoy, S. N. Dixit, R. G. Tonkyn, and M. G. White, *J. Chem. Phys.* **93**, 5345 (1990).

208. P. M. Guyon, T. Baer, and I. Nenner, *J. Chem. Phys.* **78**, 3665 (1983).
209. V. Beutel, G. L. Bhale, M. Kuhn, and W. Demtröder, *Chem. Phys. Lett.* **185**, 313 (1991).
210. V. Beutel, H.-G. Krämer, G. L. Bhale, M. Kuhn, K. Weyers, and W. Demtröder, *J. Chem. Phys.* **98**, 2699 (1993).
211. T. P. Softley, A. J. Hudson, and R. Watson, *J. Chem. Phys.* **in press.**
212. E. W. Schlag and R. D. Levine, *to be published.*
213. D. R. Inglis and E. Teller, *Astrophys. J.* **90**, 439 (1939).
214. J. Berkowitz and B. Ruscic, *J. Chem. Phys.* **93**, 1741 (1990).
215. M. J. J. Vrakking and Y. T. Lee, *J. Chem. Phys.* **102**, 8818 (1995).
216. M. J. J. Vrakking and Y. T. Lee, *J. Chem. Phys.* **102**, 8833 (1995).
217. E. Rabani, L. Ya. Baranov, R. D. Levine, and U. Even, *Chem. Phys. Lett.* **221**, 473 (1994).
218. L. Ya. Baranov, F. Remacle, and R. D. Levine, *Phys. Rev. A* **54**, 4789 (1996).
219. A. Held, L. Ya. Baranov, H. L. Selzle, and E. W. Schlag, *J. Chem. Phys.* **106**, 6848 (1997).
220. C. Y. Ng, in *The Structure, Energetics and Dynamics of Organic Molecules*, edited by T. Baer, C. Y. Ng, and I. Powis (Wiley, New York, 1996).
221. J. D. D. Martin, C. Alcaraz, and J. W. Hepburn, *J. Phys. Chem. (Y. T. Lee Festschrift)* **in press** (1997).
222. R. R. Jones, P. Fu, and T. F. Gallagher, *J. Chem. Phys.* **106**, 3578 (1997).
223. J. R. Rubbmark, M. M. Kash, M. G. Littmann, and D. Kleppner, *Phys. Rev. A* **23**, 3107 (1981).
224. J. H. M. Neijzen and A. Dönszelmann, *J. Phys. B: Atomic Mol. Phys.* **15**, L87 (1982).
225. K. H. Fung, H. L. Selzle, and E. W. Schlag, *Z. Naturforsch. A* **36**, 1257 (1981).
226. L. Ya. Baranov, R. Kris, R. D. Levine, and U. Even, *J. Chem. Phys.* **100**, 186 (1994).
227. H.-J. Dietrich, K. Müller-Dethlefs, and L. Ya. Baranov, *Phys. Rev. Lett.* **79**, 3530 (1996).
228. S. T. Pratt, *J. Chem. Phys.* **98**, 9241 (1993).
229. A. Held, H. L. Selzle, and E. W. Schlag, *J. Phys. Chem.* **100**, 15314 (1996).
230. A. Strobel, I. Fischer, J. Staecker, G. Niedner-Schatteburg, K. Müller-Dethlefs, and V. E. Bondybey, *J. Chem. Phys.* **97**, 2332 (1992).
231. P. Hobza, R. Burcl, V. Spirko, O. Dopfer, K. Müller-Dethlefs, and E. W. Schlag, *J. Chem. Phys.* **101**, 990 (1994).
232. K. Fuke, H. Yoshiuchi, K. Kaya, Y. Achiba, K. Sato, and K. Kimura, *Chem. Phys. Lett.* **108**, 179 (1984).
233. R. D. Amos and J. E. Rice, *The Cambridge Analytic Derivatives Package* **issue 4.0**, Cambridge (1997).
234. M. J. Frisch, G. W. Trucks, H. B. Schlegel, P. M. W. Gill, B. G. Johnson, M. A. Robb, J. R. Cheeseman, T. A. Keith, G. A. Petersson, J. A. Montgomery, K. Raghavachari, M. A. Al-Laham, V. G. Zakrzewski, J. V. Ortiz, J. B. Foresman, J. Cioslowski, B. B. Stefanov, A. Nanayakkara, M. Challacombe, C. Y. Peng, P. Y. Ayala, W. Chen, M. W. Wong, J. L. Andres, E. S. Replogle, R. Gomperts, R. L. Martin, D. J. Fox, J. S. Binkley, D. J. Defrees, J. Baker, J. P. Stewart, M. Head-Gordon, C. Gonzalez, and J. A. Pople, *Gaussian Inc.* **Pittsburgh, PA** (1995).
235. N. J. van Druten and H. G. Muller, *Phys. Rev. A* **52**, 3047 (1995).
236. C. E. Alt, W. G. Scherzer, H. L. Selzle, E. W. Schlag, L. Ya. Baranov, and R. D. Levine, *J. Phys. Chem.* **99**, 1660 (1995).

237. R. Rednall and T. P. Softley, *in preparation*
238. ZEKE home page, *http://www.chemie.tu-muenchen.de/zeke* (maintained by Dr W. J. Knott, Institut für physikalische Chemie, Technische Universität München).
239. M. Karplus and J. N. Kushik, *Macromolecules* **14**, 325 (1981).
240. M. Karplus and J. A. McCammon, *CRC Crit. Rev. Biochem.* **9**, 293 (1981).
241. M. Karplus and J. A. McCammon, *Fed. Eur. Biochem. Soc. Lett.* **131**, 34 (1981).
242. M. Karplus and J. A. McCammon, *Ann. Rev. Biochem.* **52**, 263 (1983).
243. M. Karplus and J. A. McCammon, *Scient. Am.* **4**, 42 (1986).
244. A. H. Zewail, in *Femtosecond Chemistry*, edited by J. Manz and L. Wöste (VCH Weinheim, New York, 1995).
245. T. Baumert, R. Thalweiser, V. Weiss, and G. Gerber, in *Femtosecond Chemistry*, edited by J. Manz and L. Wöste (VCH Weinheim, New York, 1995).
246. Y. R. Shen, *Ann. Rev. Phys. Chem.* **40**, 327 (1989).
247. T. Oka, in *The Spectroscopy of Molecular Ions: Proceedings of a Royal Society Discussion*, edited by A. Carrington and B. A. Thrush (Royal Society, London, 1988).
248. R. J. Saykally and R. C. Woods, *Ann. Rev. Phys. Chem.* **32**, 403 (1981).
249. C.S. Gudeman and R. J. Saykally, *Ann. Rev. Phys. Chem.* **35**, 387 (1984).
250. K. F. Willey, C. S. Yeh, and M. A. Duncan, *Chem. Phys. Lett.* **211**, 156 (1993).
251. J. E. Harrington and J. C. Weisshaar, *J. Chem. Phys.* **93**, 854 (1990).
252. W. Kong, D. Rodgers, J. Hepburn, K. Wang, and V. McKoy, *J. Chem. Phys.* **99**, 3159 (1993).
253. F. Merkt, S. R. Mackenzie, R. J. Rednall, and T. P. Softley, *J. Chem. Phys.* **99**, 8430 (1993).
254. I. Fischer, A. Lochschmidt, A. Strobel, G. Niedner-Schatteburg, K. Müller-Dethlefs, and V. E. Bondybey, *Chem. Phys. Lett.* **202**, 542 (1993).
255. A. Strobel, A. Lochschmidt, I. Fischer, G. Niedner-Schatteburg, and V. E. Bondybey, *J. Chem. Phys.* **99**, 733 (1993).
256. A. Strobel, I. Fischer, A. Lochschmidt, K. Müller-Dethlefs, and V. E. Bondybey, *J. Phys. Chem.* **98**, 2024 (1994).
257. C.-W. Hsu and C. Y. Ng, *J. Chem. Phys.* **101**, 5596 (1994).
258. S. T. Pratt, P. M. Dehmer, and J. L. Dehmer, *J. Chem. Phys.* **99**, 6233 (1993).
259. N. Knoblauch, A. Strobel, I. Fischer, and V. E. Bondybey, *J. Chem. Phys.* **103**, 5417 (1995).
260. L. Zhu and P. Johnson, *J. Chem. Phys.* **99**, 2322 (1993).
261. H. Ozeki, M. C. R. Cockett, K. Okuyama, M. Takahashi, and K. Kimura, *J. Phys. Chem.* **99**, 8608 (1995).
262. T. G. Wright, S. I. Panov, and T. A. Miller, *J. Chem. Phys.* **102**, 4793 (1995).
263. G. Lembach and B. Brutschy, *Chem. Phys. Lett.* **237**, 421 (1997).
264. G. Reiser, D. Rieger, T. G. Wright, K. Müller-Dethlefs, and E. W. Schlag, *J. Phys. Chem.* **97**, 4335 (1993).
265. K. Müller-Dethlefs and E. W. Schlag, *Ann. Rev. Phys. Chem.* **42**, 109 (1991).
266. H. Shinohara, S. I. Sato, and K. Kimura, *J. Phys. Chem. A* **101**, 6736 (1997).
267. R. G. Neuhauser, K. Siglow, and H. J. Neusser, *J. Chem. Phys.* **106**, 896 (1997).
268. H. Krause and H. J. Neusser, *Chem. Phys. Lett.* **213**, 603 (1993).
269. H. J. Neusser and H. Krause, *Chem. Rev.* **84**, 1829 (1994).
270. O. Dopfer and K. Müller-Dethlefs, *J. Chem. Phys.* **101**, 8508 (1994).

271. O. Dopfer, M. Melf, and K. Müller-Dethlefs, *Chem. Phys.* **207**, 437 (1996).
272. T. G. Wright, E. Cordes, O. Dopfer, and K. Müller-Dethlefs, *J. Chem. Soc. Faraday Trans.* **89**, 1609 (1993).
273. E. Cordes, O. Dopfer, T. G. Wright, and K. Müller-Dethlefs, *J. Phys. Chem.* **97**, 7471 (1993).
274. O. Dopfer, T. G. Wright, E. Cordes, and K. Müller-Dethlefs, *J. Am. Chem. Soc.* **116**, 5880 (1994).
275. M. C. R. Cockett, M. Takahashi, K. Okuyama, and K. Kimura, *Chem. Phys. Lett.* **187**, 250 (1991).
276. M. Takahashi, H. Ozeki, and K. Kimura, *J. Chem. Phys.* **96**, 6399 (1992).
277. J. M. Smith, X. Zhang, and J. L. Knee, *J. Chem. Phys.* **99**, 2550 (1993).
278. K.-T. Lu and J. C. Weisshaar, *J. Chem. Phys.* **99**, 4247 (1993).
279. M. G. H. Boogaarts, I. Holleman, R. T. Jongma, D. H. Parker, G. Meijer, and U. Even, *J. Chem. Phys.* **104**, 4357 (1996).
280. M. Araki, S. I. Sato, and K. Kimura, *J. Phys. Chem.* **100**, 10542 (1996).
281. R. A. Walker, E. C. Richard, K.-T. Lu, and J. C. Weisshaar, *J. Phys. Chem.* **99**, 12422 (1995).
282. R. A. Walker, E. C. Richard, and J. C. Weisshaar, *J. Phys. Chem.* **100**, 7333 (1996).
283. H. Ozeki, M. Takahashi, K. Okuyama, and K. Kimura, *J. Chem. Phys.* **99**, 56 (1993).
284. K. Takazawa, M. Fujii, and M. Ito, *J. Chem. Phys.* **99**, 3205 (1993).
285. K. T. Lu, G. C. Eiden, and J. C. Weisshaar, *J. Phys. Chem.* **96**, 9742 (1992).
286. J. M. Dyke, H. Ozeki, M. Takahashi, M. C. R. Cockett, and K. Kimura, *J. Chem. Phys.* **97**, 8926 (1992).
287. T. Vondrak, S. I. Sato, and K. Kimura, *J. Phys. Chem.* **101**, 2384 (1997).
288. M. C. R. Cockett, K. Okuyama, and K. Kimura, *J. Chem. Phys.* **97**, 4679 (1992).
289. M. Takahashi and K. Kimura, *J. Chem. Phys.* **97**, 2920 (1992).
290. X. Song, R. R. Davidson, S. R. Gwaltney, and J. P. Reilly, *J. Chem. Phys.* **100**, 5411 (1994).
291. M. C. R. Cockett, H. Ozeki, K. Okuyama, and K. Kimura, *J. Chem. Phys.* **98**, 7763 (1993).
292. T. Vondrak, S. Sato, and K. Kimura, *Chem. Phys. Lett.* **261**, 481 (1996).
293. C. Lakshminarayan, J. M. Smith, and J. L. Knee, *Chem. Phys. Lett.* **182**, 656 (1991).
294. Th. L. Grebner and H. J. Neusser, *Chem. Phys. Lett.* **245**, 578 (1995).
295. H. J. Neusser, H. Krause, T. L. Grebner, R. Sußmann, and R. Neuhauser, *Proc. SPIE* **258**, 91 (1995).
296. O. Dopfer, G. Lembach, T. G. Wright, and K. Müller-Dethlefs, *J. Chem. Phys.* **98**, 1933 (1993).
297. K. Okuyamy, M. C. R. Cockett, and K. Kimura, *J. Chem. Phys.* **97**, 1649 (1992).
298. M. C. R. Cockett and K. Kimura, *J. Chem. Phys.* **100**, 3429 (1994).
299. J. M. Smith and J. L. Knee, *Laser Chem.* **14**, 131 (1994).
300. J. M. Smith and J. L. Knee, *J. Chem. Phys.* **99**, 38 (1993).
301. N. P. Wales, W. J. Buma, C. A. de Lange, H. Lefebvre-Brion, K. Wang, and V. McKoy, *J. Chem. Phys.* **104**, 4911 (1996).
302. R. T. Wiedmann and M. G. White, *J. Chem. Phys.* **102**, 5141 (1995).
303. R. G. Tonkyn, R. T. Wiedmann, and M. G. White, *J. Chem. Phys.* **96**, 3696 (1992).
304. A. Mank, D. Rodgers, and J. W. Hepburn, *Chem. Phys. Lett.* **219**, 169 (1994).

305. S. T. Pratt, *J. Chem. Phys.* **101**, 8302 (1994).
306. C.-W. Hsu, D. P. Baldwin, C.-L. Liao, and C. Y. Ng, *J. Chem. Phys.* **100**, 8047 (1994).
307. R. G. Tonkyn, R. Wiedmann, E. R. Grant, and M. G. White, *J. Chem. Phys.* **95**, 7033 (1991).
308. R. T. Wiedmann and M. G. White, *Proc. SPIE* **1638**, 273 (1992).
309. M. C. R. Cockett, J. G. Goode, K. P. Lawley, and R. J. Donovan, *J. Chem. Phys.* **102**, 5226 (1995).
310. J. G. Goode, M. C. R. Cockett, K. P. Lawley, and R. J. Donovan, *Chem. Phys. Lett.* **231**, 521 (1994).
311. R. Signorell, H. Palm, and F. Merkt, *J. Chem. Phys.* **106**, 6523 (1997).
312. M. Takahashi, *J. Chem. Phys.* **96**, 2594 (1991).
313. G. P. Bryant, Y. N. Jiang, M. Martin, and E. R. Grant, *J. Chem. Phys.* **101**, 7199 (1994).
314. F. Merkt and T. P. Softley, *Phys. Rev. A* **46**, 302 (1992).
315. R. T. Wiedmann, E. R. Grant, R. G. Tonkyn, and M. G. White, *J. Chem. Phys.* **95**, 746 (1991).
316. A. Strobel, N. Knoblauch, J. Agreiter, A. M. Smith, G. Niedner-Schatteburg, and V. E. Bondybey, *J. Phys. Chem.* **99**, 872 (1995).
317. R. Thalweiser, S. Vogler, and G. Gerber, *Proc. SPIE* **1858**, 196 (1993).
318. D. A. Rodham and G. A. Blake, *Chem. Phys. Lett.* **264**, 522 (1997).
319. D.-Sh. Yang, M. Z. Zgierski, A. Bérces, P. A. Hackett, P.-N. Roy, A. Martinez, T. Carrington Jr, D. R. Salahub, R. Fournier, T. Pang, and Ch. Chen, *J. Chem. Phys.* **105**, 10663 (1996).
320. D. S. Yang, M. Z. Zgierski, A. Bérces, P. A. Hackett, A. Martinez, and D. R. Salahub, *Chem. Phys. Lett.* **277**, 71 (1997).
321. D. S. Yang, M. Z. Zgierski, D. M. Rayner, P. A. Hackett, A. Martinez, D. R. Salahub, P.-N. Roy, and T. Carrington Jr, *J. Chem. Phys.* **103**, 5335 (1995).
322. J. Harrington and J. C. Weisshaar, *J. Chem. Phys.* **97**, 2809 (1992).
323. D. S. Yang, A. M. James, D. M. Rayner, and P. A. Hackett, *J. Chem. Phys.* **102**, 3129 (1995).
324. D. S. Yang, A. M. James, D. M. Rayner, and P. A. Hackett, *Chem. Phys. Lett.* **231**, 177 (1994).
325. D.-S. Yang, B. Simard, P. A. Hackett, A. Bréces, and M. Z. Zgierski, *Int. J. Mass Spectrom. Ion Processes* **159**, 65 (1996).
326. I. Yourshaw, Y. Zhao, and D. M. Neumark, *J. Chem. Phys.* **105**, 351 (1996).
327. G. F. Ganteför, D. M. Cox, and A. Kaldor, *J. Chem. Phys.* **96**, 4102 (1992).
328. C. C. Arnold, D. M. Neumark, D. M. Cyr, and M. A. Johnson, *J. Phys. Chem.* **99**, 1633 (1995).
329. G. Drechsler, U. Boesl, C. Bäßmann, and E. W. Schlag, *J. Chem. Phys.* **107**, 2284 (1997).
330. T. N. Kitsopoulos, I. M. Waller, J. G. Loeser, and D. M. Neumark, *Chem. Phys. Lett.* **159**, 300 (1989).
331. J. R. Smith, J. B. Kim, and W. C. Lineberger, *Phys. Rev. A* **55**, 2036 (1997).
332. C. C. Arnold and D. M. Neumark, *J. Chem. Phys.* **99**, 3353 (1993).

Subject index

Author index

Page number A1 indicates there is a citation in the Appendix.

Printed in the United States
By Bookmasters